An Unauthorized Guide to Mobil Collectibles:

Chasing the Red Horse

Rob Bender
Tammy Cannoy-Bender

4880 Lower Valley Road, Atglen, PA 19310 USA

Dedication

We wish to dedicate this book to my step-grandfather, Clinton Borton, who started selling and delivering Mobil products at the age of 16. He did this, taking time out only for World War II, until he passed away in 1970. He is the reason for our obsession with collecting Mobil petrolania.

Library of Congress Catalog Card Number:

Book design by: Anne Davidsen
Type set in Futura Heavy/ Humanist 521

ISBN: 0-7643-0782-7
Printed in China
1 2 3 4

Published by Schiffer Publishing Ltd.
4880 Lower Valley Road
Atglen, PA 19310
Phone: (610) 593-1777; Fax: (610) 593-2002
E-mail: Schifferbk@aol.com

This book may be purchased from the publisher.
Please visit out web site catalog at **www.schifferbooks.com**
Include $3.95 for shipping.
Please try your bookstore first.
We are interested in hearing from authors
with book ideas on related subjects.
You may write for a free catalog.

In Europe, Schiffer books are distributed by
Bushwood Books
6 Marksbury Rd.
Kew Gardens
Surrey TW9 4JF England
Phone: 44 (0)181 392-8585; Fax: 44 (0)181 392-9876
E-mail: Bushwd@aol.com

Table of Contents

Please note that the prices enclosed are merely a guide. The location of purchase, the rarity, and condition of the item will have a varying effect on the price. Collectors should use their own judgement.

Acknowledgements

We would like to give special thanks to the following people for their help and support thorough out the process of making this book.
Bob & Ellen Bender, Cazenovia, WI
Bill & Sal Borton, McFarland, WI
Gary & Connie Cannoy, Muscoda, WI
George & Mimi Deel, Blue Springs, MO (816-229-7223)
Robert McCalla, "White Eagle Antique Mall", Augusta, KS
Bernice & Ward McDonald, Richland Center, WI
Mike Orr, Dodgeville, WI
Peter Palermo, Charlton, NY, for taking pictures of his collection (518-882-9893)
Dale Weien, "Dale Weien Auction Enterprises" Ottawa, KS

We would also like to thank all of the people along the way in the making of this book that have given us advice and direction.

Foreword

Today's Mobil is and was made up of several oil companies that were bought or absorbed over the years. This book is going to try to show a little history of Mobil as it evolved.

We are going to try to show products from all or as many of these companies as possible, which include names like White Eagle, Vacuum Oil, Socony, General Petroleum, Magnolia, Lubrite, Wadhams, Gilmore and White Star.

Collecting petroleum memorabilia is an ever-growing hobby enjoyed by many people of different lifestyles. Some people collect all brands and some collect just one brand. We are one of those who collect any thing to do with Mobil Oil. This book is for everybody who, like us, can't stop chasing the Pegasus.

Please enjoy reading as much as we did writing this book.

A Brief History

In 1866, Vacuum Oil Co. was incorporated in Rochester New York. Its most significant product was patented in 1869, Gargoyle "600-W" Steam Cylinder Oil. This oil, which is still in demand today, made it possible to build bigger and better machines to increase output.

In 1879 Standard Oil Co. purchased a three-quarter interest in Vacuum Oil Co. In 1882, the Standard Oil Trust was formed in New York. Later that year, Standard Oil Co. of New York was formed, or as it was also known, Socony. In 1906, the U.S. Government filed an antitrust suit which caused Standard Oil to break up into 33 smaller companies in 1911. Socony and Vacuum were two of these. 1911 was also the first year that oil companies sold more gasoline than kerosene.

Over the next few years Socony & Vacuum expanded to keep up with demand and to expand into new areas. In 1925 Socony acquired complete ownership of Magnolia Petroleum Co., which included holdings in Beaumont, Texas. In 1926 Socony purchased General Petroleum Corp of California. To establish holdings throughout the country, Socony purchased White Eagle Oil & Refining Co. which was located in the midwestern United States. While Socony was expanding into several new market areas, Vacuum was also expanding. In 1929 Vacuum acquired Lubrite Refining Co. in St. Louis and in 1930 purchased Wadhams Oil Corp in Milwaukee. Also in 1930 Vacuum acquired White Star Refining Co., which had two refineries, a series of bulk plants, and 1500 filling stations in Michigan, Indiana and Ohio.

In 1931 Socony and Vacuum were combined. Shortly after this merger the Pegasus, or Flying Red Horse, was adopted as its trademark in the United States. The Flying Red Horse had been used as early as 1911 overseas by a former company of Mobil's in South Africa. The Pegasus was white at this time. It did not become red until Mobil Sekiyu in Japan used it shortly thereafter.

To help solve its overseas distribution problems, Socony-Vacuum became equal partners with Standard Oil of New Jersey in 1935 to form Standard-Vacuum Oil Co. or Stan-Vac.

In 1934 Socony-Vacuum Corporation went through another name change. This time it was Socony-Vacuum Oil Company Incorporated. This is also when the word "Mobil" was registered as a trademark. Mobil was coined from the two words Mobiloil and Mobilgas. Mobiloil is the oldest of these two words. It was first used in England in 1899. The Mobiloil trademark was registered in the United States in January of 1920.

When war came in 1939, Socony-Vacuum was involved. The company's tanker, the SS *Emidio*, was the first United States tanker sunk by the enemy. During World War II, Socony-Vacuum lost 432 men and 32 ships. They also temporarily lost their refineries in France and Italy. The refinery in Gravenchon France was the first. On June 9, 1940, the French High Command cabled the refinery to "BURN ALL YOUR RESERVES." The smoke from this fire could be seen for 30 miles. By 1946 Gravenchon had been rebuilt and online. During the war other Socony-Vacuum affiliates also suffered similar experiences. Some actually ceased to exist during this time. The next few years were spent rebuilding and expanding. The Mobil name was now becoming familiar around the globe. In 1955 Socony-Vacuum decided to take advantage of the value of recognition of the word "Mobil"; there was another name change to Socony-Mobil Oil Company Inc. In 1959 Socony-Mobil reorganized so that all the companies acquired over the years, like Magnolia, Gilmore, and General Petroleum were merged into one parent company, Socony-Mobil. Over the next few years Mobil expanded into petrochemicals. Also, Stan-Vac was disassembled and Socony-Mobil's interests around the world became more united. In 1966 a new name, Mobil Oil Corporation, was approved by stockholders. The word "Mobil" with its red "o" now became the key symbol on the service stations' signs. The Pegasus, which had been the symbol for over 35 years, moved to the sides of the buildings. Since this change Mobil has kept growing and changing with the times. Maybe in 30 to 50 years the signs and products of today's Mobil will be as collected and valuable as the signs from Mobil 30 to 50 years ago.

I. Cans

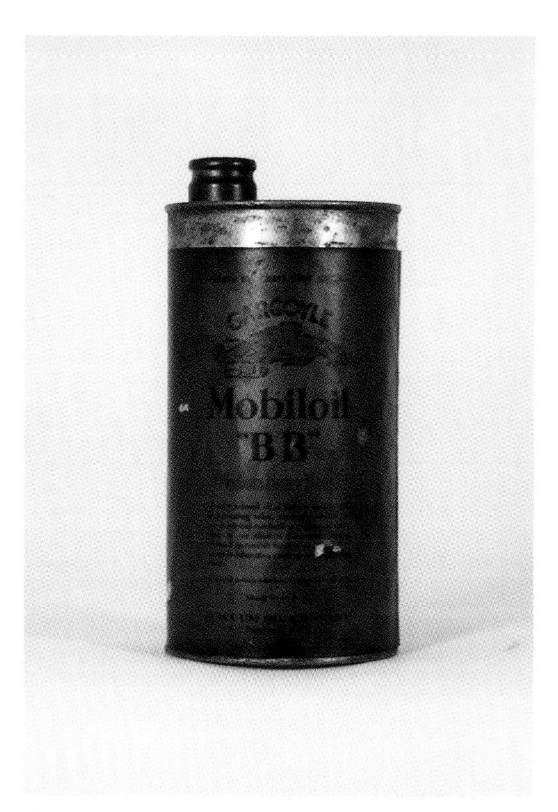

Gargoyle, Vacuum Oil Co., paper label, one-quart Mobiloil "BB". $150–200

Gargoyle, Vacuum Oil Co., paper label, one-quart Arctic Oil "C". $150–200

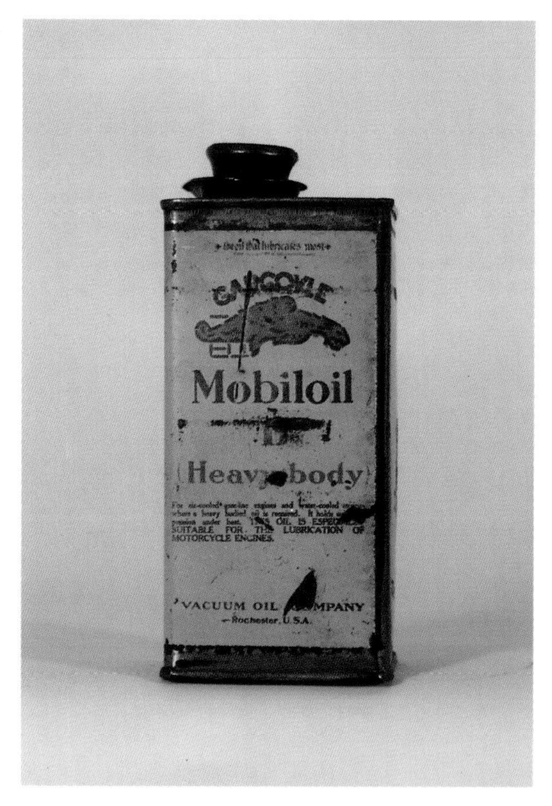

Gargoyle, Vacuum Oil Co., one-quart square Mobiloil "D". $125–150

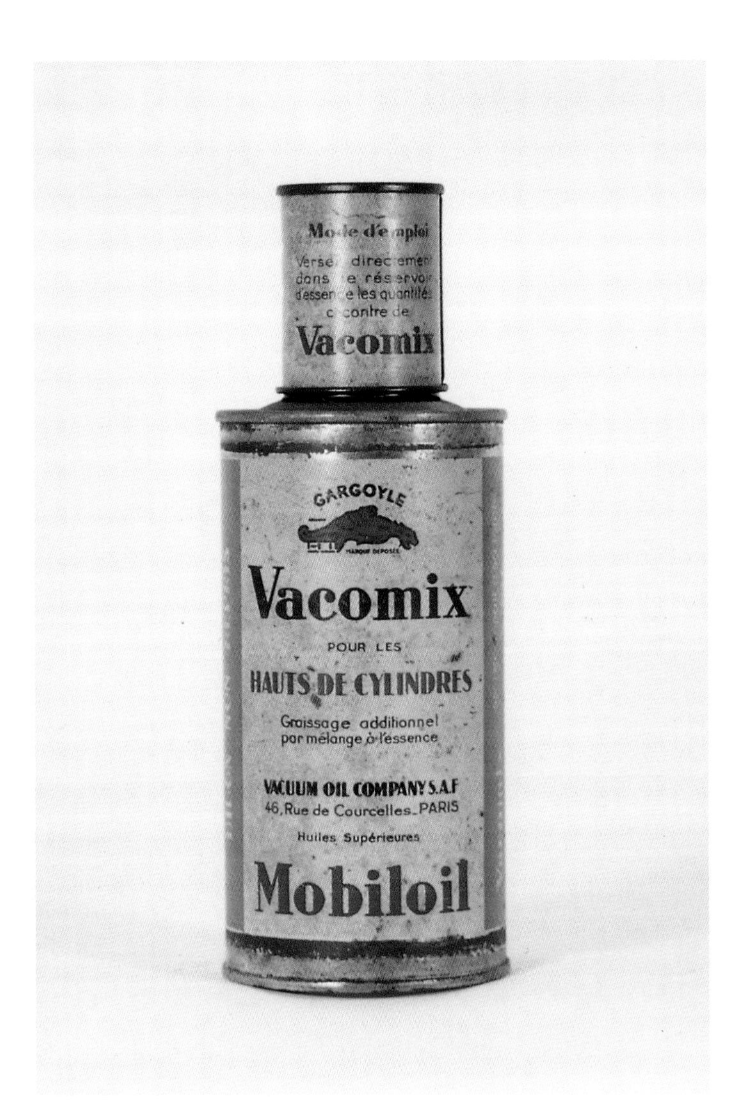

Gargoyle, Vacuum Oil Co., one-pint can with measure cup on top, foreign can from France, "Vacomix". $400.

Top of can from France.

This is the top of the Gargoyle one-quart oil cans on the following page.

Gargoyle, Vacuum Oil Co., one-quart Mobiloil "A". $150–200.

Side view of Gargoyle, Mobiloil "A". "FAIR RETAIL PRICE 35c. THREE FOR $1.00, (Slightly higher in Southwestern, Mountain and Pacific Coast States) Net Content one quart".

Gargoyle, Vacuum Oil Co., one-quart Mobiloil "B". $150–200

Gargoyle, Vacuum Oil Co., Mobiloil "E", "Especially Recommended for Ford Cars". $200–225.

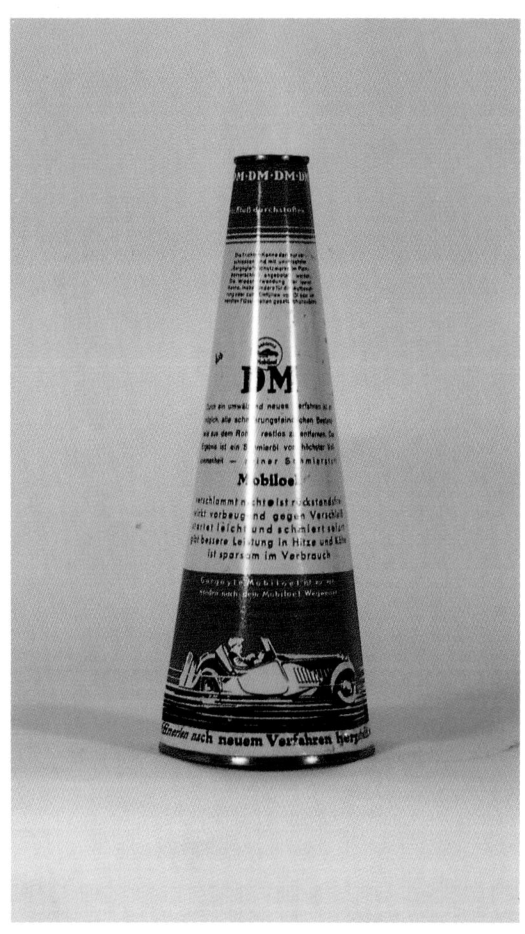

Gargoyle, Vacuum Oil Co., foreign cone-shaped can from Germany, Mobiloel DM with a motorcycle on the front and a car on the back. $600.

Gargoyle, Socony-Vacuum Oil Co., one-quart can Arctic. $175–200

Socony-Vacuum Oil Co. Inc., steel
one-quart can. $45–60.

Gargoyle, Socony-Vacuum Oil Co., one-quart can Mobiloil "A". $275–200

Gargoyle, Socony-Vacuum Oil Co., one-quart can
Arctic Special, label on the top of the can. $50–75

The top of the Gargoyle Arctic Special can.

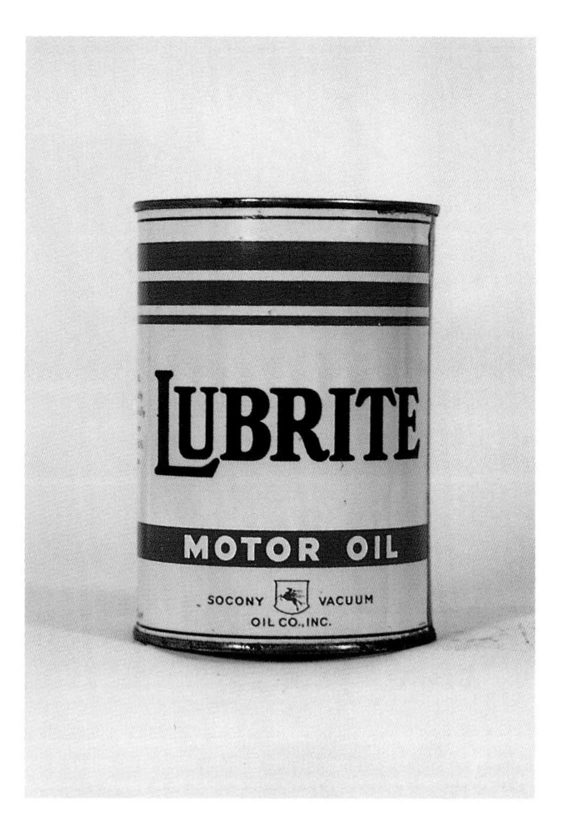

Socony-Vacuum Oil Co. Inc., steel one-quart
Lubrite Motor Oil, white can. $50–75

Socony-Vacuum Oil Co. Inc., steel one-quart
Standard Lubrite Motor Oil can. $35–50

Socony-Vacuum Oil Co. Inc, steel one-quart
Lubrite Motor Oil can. $35–50

Socony-Vacuum Oil Co., steel one-quart
Mobiloil Aero Gray Band can. $35–50

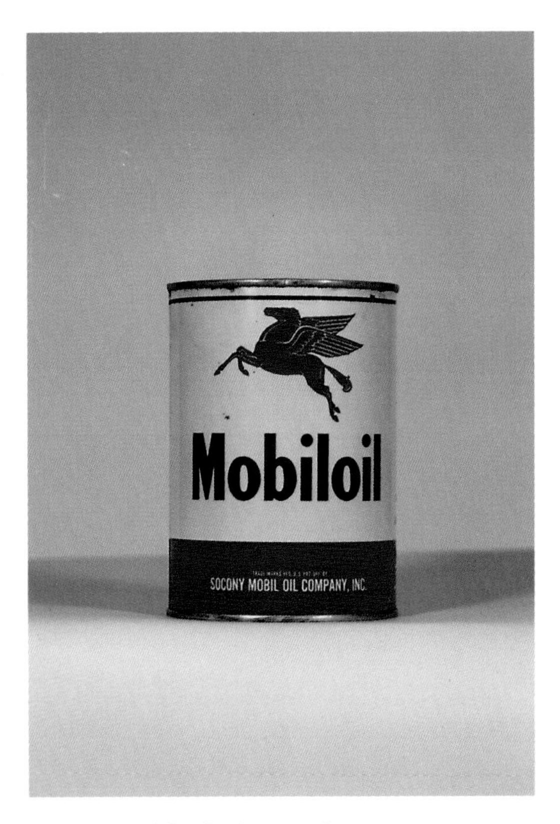

Socony Mobil Oil Co. Inc., steel one-quart
can. $35–50

Socony-Vacuum Oil Co. Inc., steel one-quart
Mobiloil Aero Red Band can. $35–50

Socony Mobil Oil Co. Inc., steel one-quart can. $35–50

Socony Mobil Oil Co. Inc., steel one-quart Lubrite Motor Oil can. $35–50

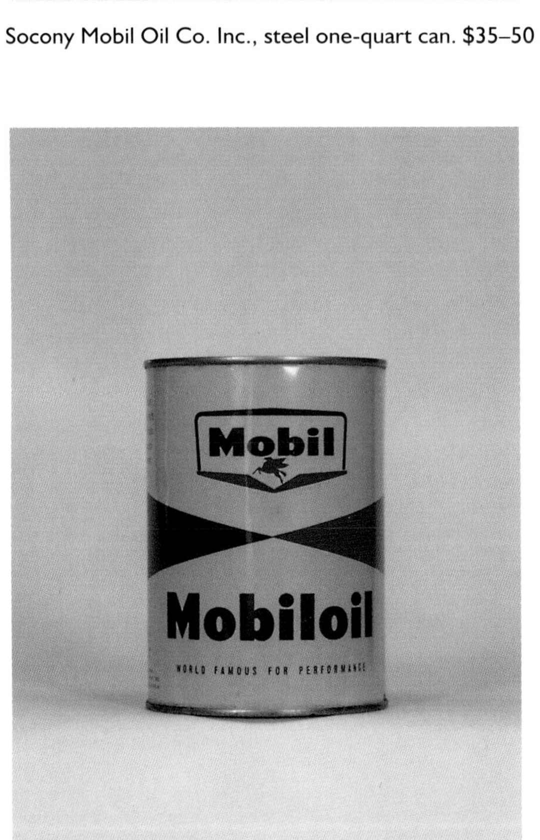

Socony Mobil Oil Co. Inc., steel one-quart Mobiloil can. $35–50

Socony Mobil Oil Co. Inc., steel one-quart Mobiloil can. $35–50

Socony Mobil Oil Co. Inc., composite one-quart Mobiloil can. $35–50

Mobil Oil Corp., composite one-quart Mobil Heavy Duty can. $12–16

Socony Mobil Oil Co. Inc., steel one-quart Mobilfluid Automatic Transmission Fluid can. $35–50

Mobil Oil Corp., composite one-quart Mobil Regular can. $12–16

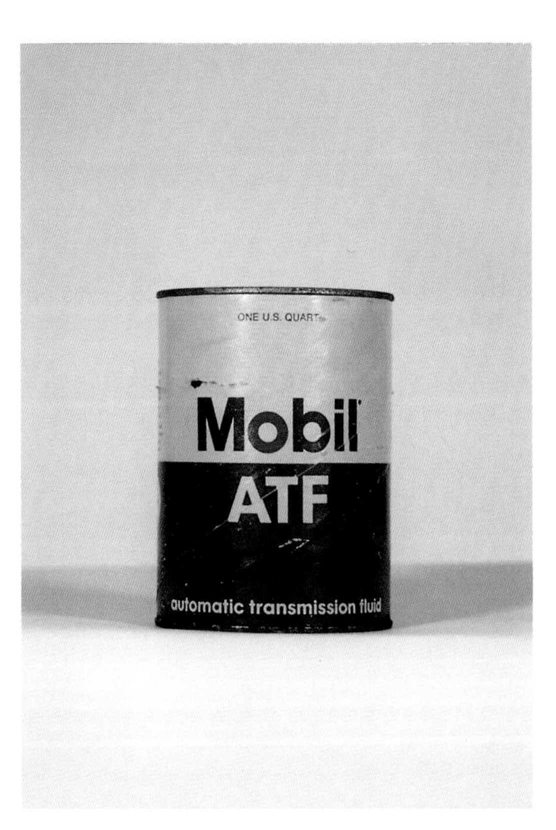

Mobil Oil Corp., composite one-quart Mobil ATF can. $12–16

Mobil Oil Corp., composite one-quart Mobil HD 30 can. $12–16

Mobil Oil Corp., composite one-quart Mobil Special 10w-30 can. $12–16

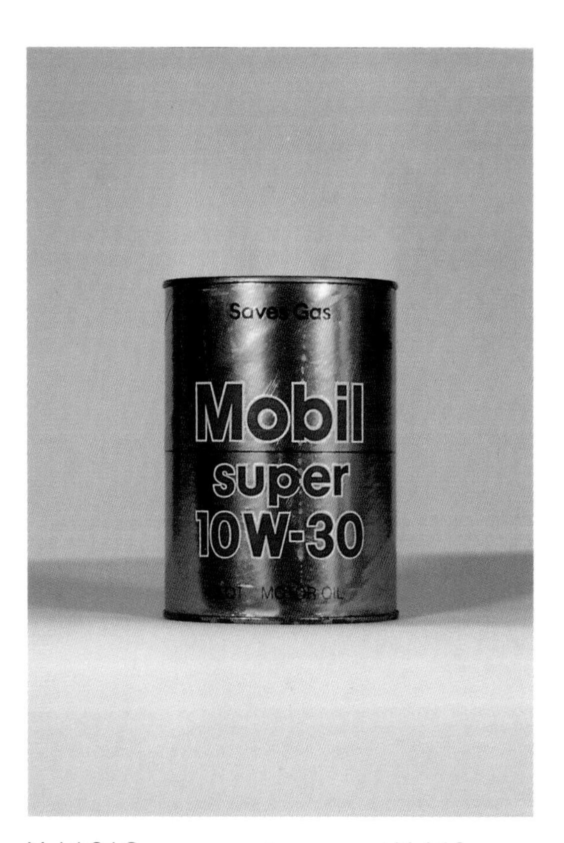

Mobil Oil Corp., composite one-quart Mobil Super 10w-30 can. $6–10

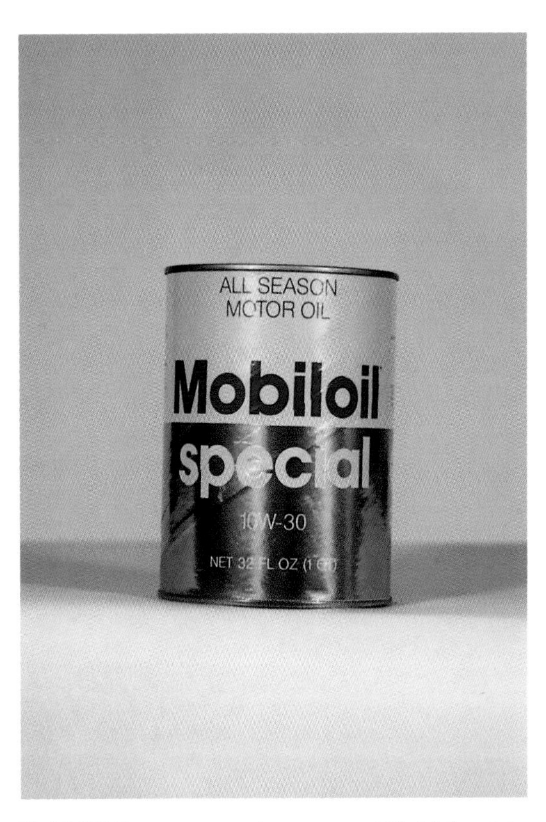

Mobil Oil Corp., composite one-quart Mobil Special 10w-30 can. $6–10

Mobil Oil Corp., composite one-quart Mobil Delvac 1200 super can. $6–10

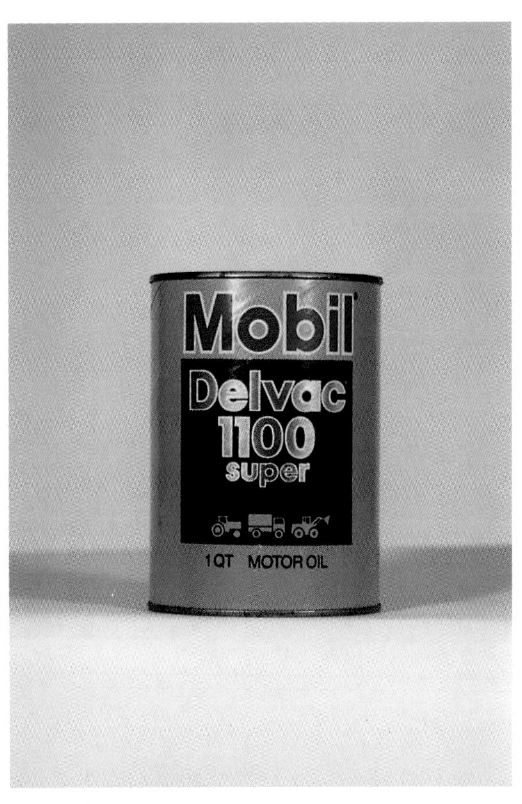

Mobil Oil Corp., composite one-quart Mobil Delvac 1100 super can. $6–10

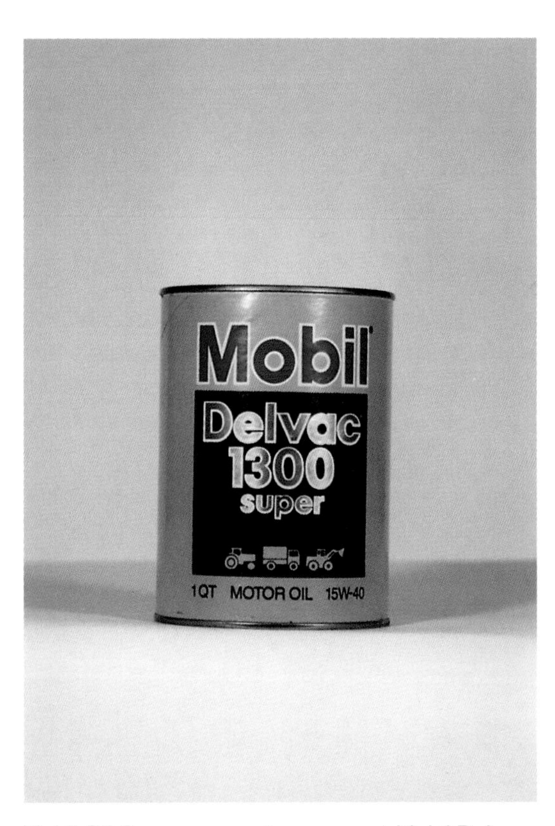

Mobil Oil Corp., composite one-quart Mobil Delvac 1300 super can. $6–10

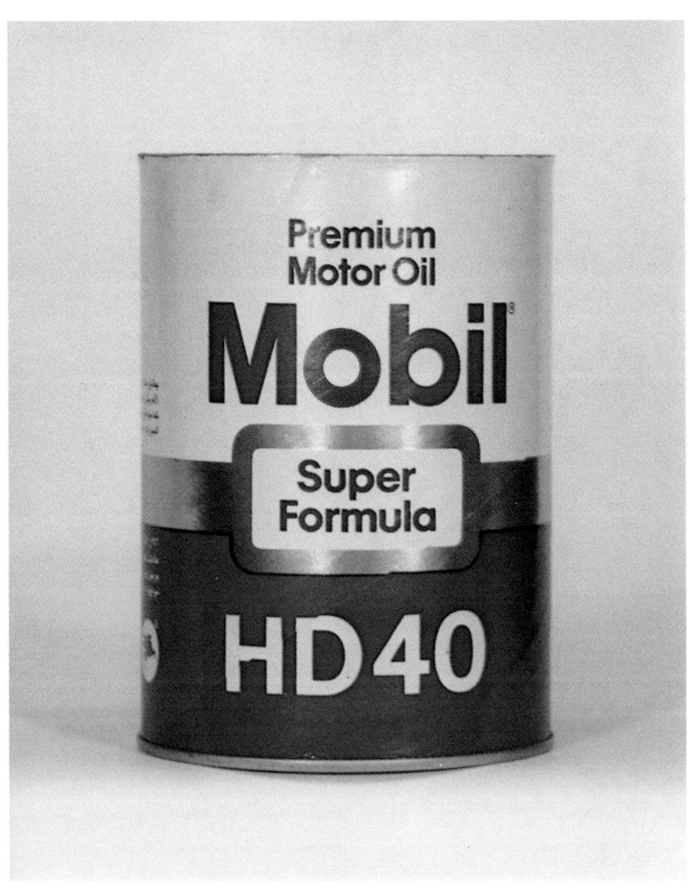

Mobil Oil Corp., composite one-quart Mobil Super Formula HD 40 in Arabic on the back side. $14–18

Back side of the can

Mobil Oil Corp., steel one-quart Mobil Jet Oil II can. $14–18

Mobil Oil Corp., steel one-quart Mobil 1 100% synthetic motor oil can. $20–25

Gargoyle, Vacuum Oil Co., German rectangular one quart Mobiloil “D” can. $500

White Eagle Oil & Refining Co., half-gallon rectangular Motor Oil can. $80

White Eagle Oil & Refining Co., half-gallon rectangular Keynoil Motor Oil can. $115

Wadhams Oil Co., half-gallon rectangular Tempered Motor Oil can.
$125–140

Standard Oil Co. of New York, rectangular gallon-sized Socony Motor Oil can. $100

Socony-Vacuum Oil Co., Inc. 2-gallon Standard Lubrite Light Hvy can. $25–35

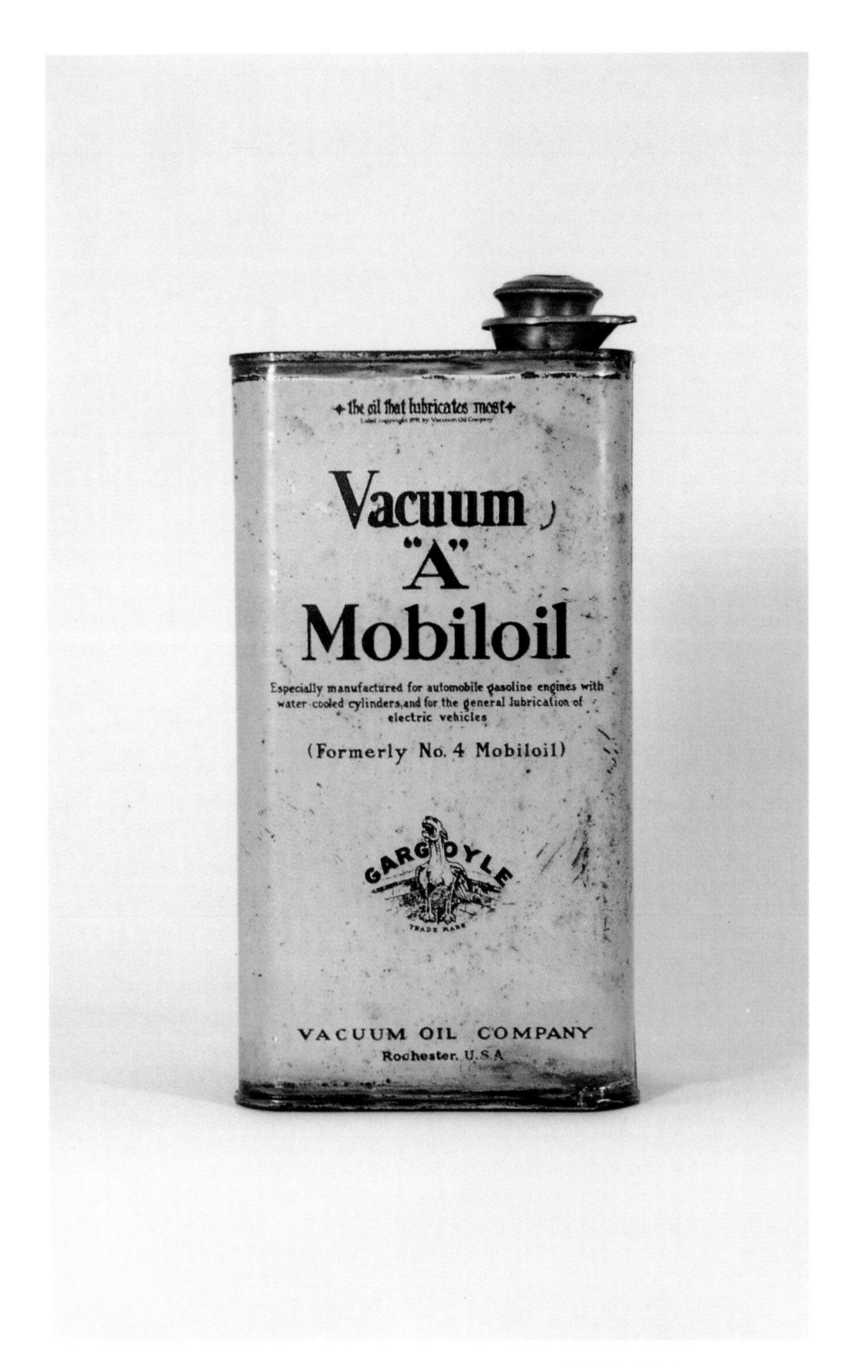

Gargoyle Vacuum Oil Co., square gallon-sized Vacuum "A" Mobiloil can. This can has the gargoyle graphic looking at the customer and has a copyright date of 1891. $475–500

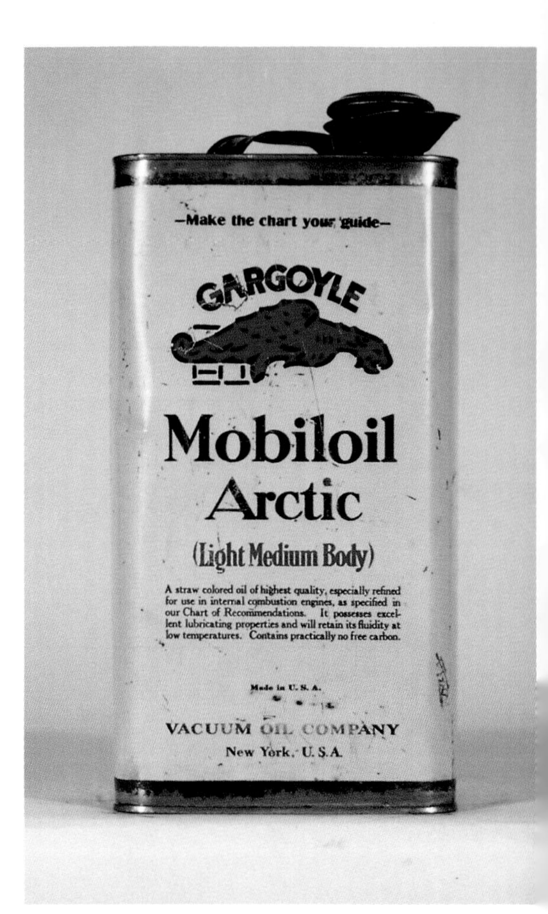

Gargoyle, Vacuum Oil Co., square gallon-sized Mobiloil Arctic (Light Medium Body) can. $150–175

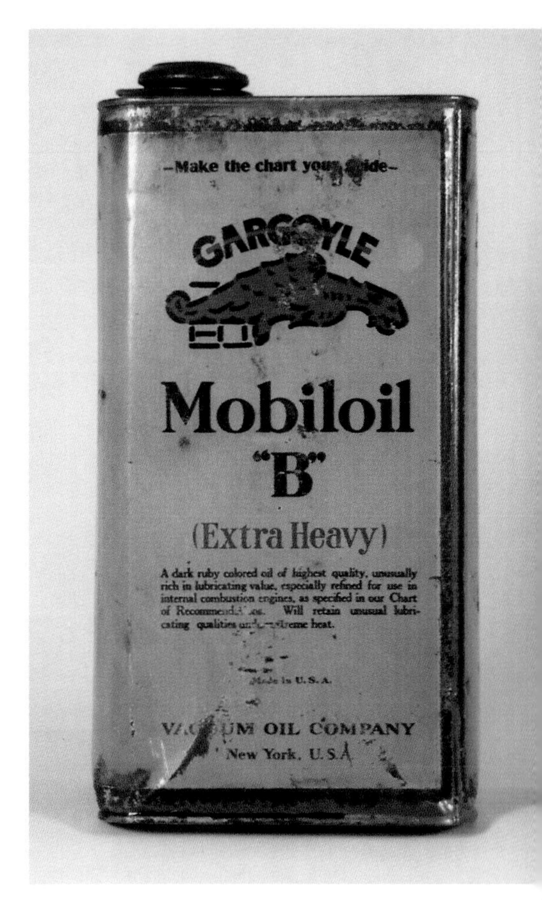

Gargoyle, Vacuum Oil Co., square gallon-sized Mobi
"B" (Extra Heavy) can. $150–175

Gargoyle, Vacuum Oil Co., square gallon-sized Mobiloil "E" can. 'Especially Recommended for Ford Cars". $175–200

Gargoyle, Vacuum Oil Co., square gallon-sized Mobiloil "BB" (Medium Heavy Body) can. $150–175

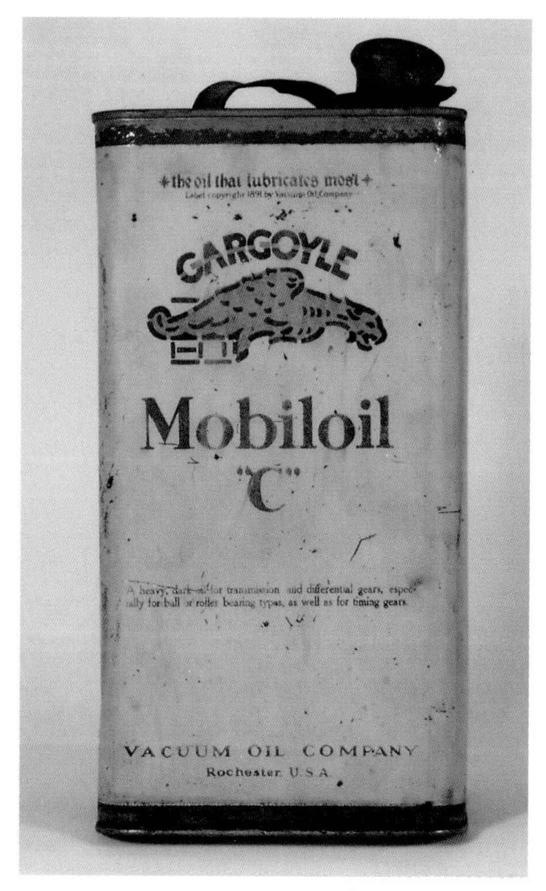

Gargoyle, Vacuum Oil Co., square gallon-sized Mobiloil "C" can. $150–175

Gargoyle, Socony-Vacuum Oil Co., square gallon-sized Mobiloil Arctic Special in Arabic. $150–175

Gargoyle, Socony-Vacuum Oil Co., square gallon-sized Mobiloil Arctic Special in Japanese. $150–175

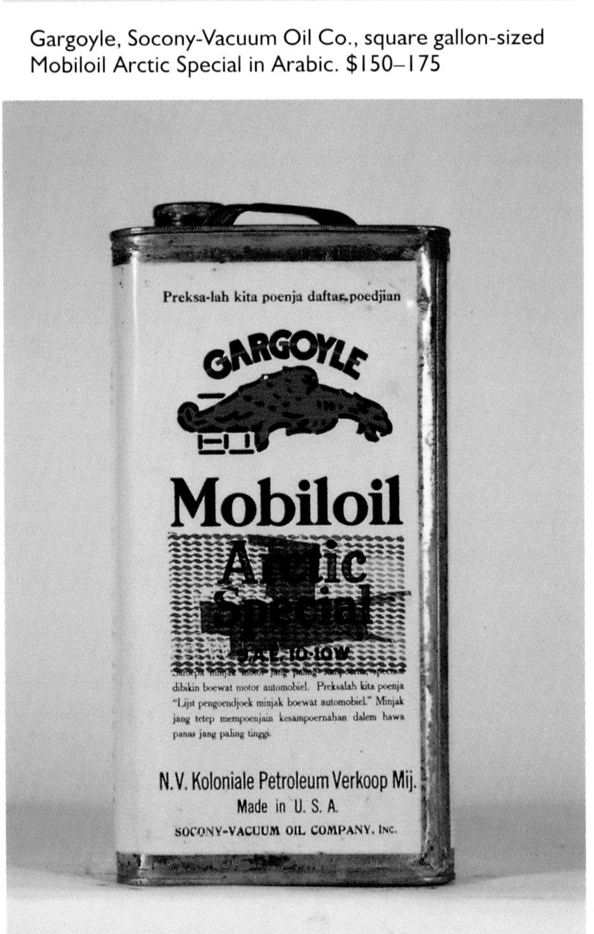

Gargoyle, Socony-Vacuum Oil Co., square gallon-sized Mobiloil Arctic Special in Swedish. $150–175

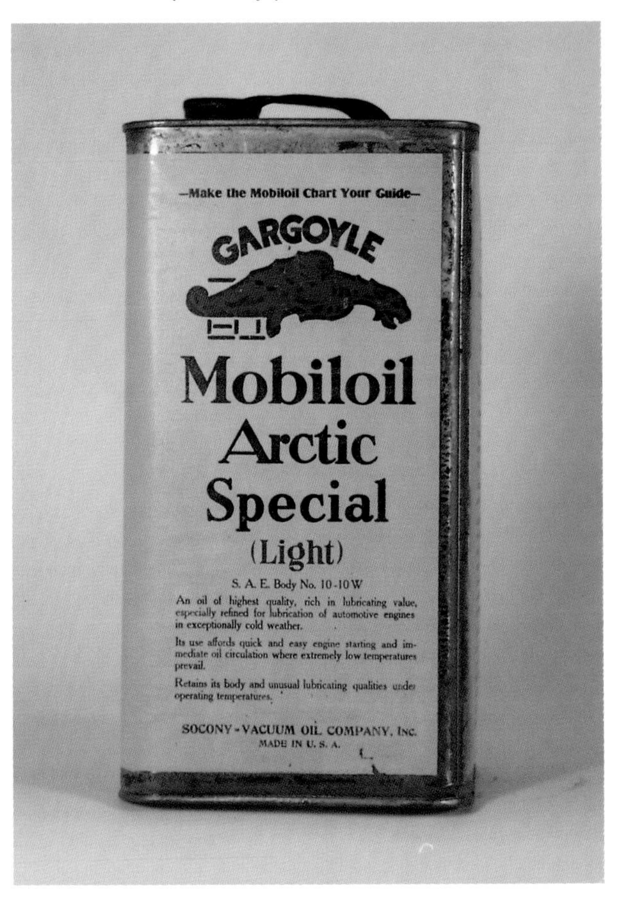

Gargoyle, Socony-Vacuum Oil Co., square gallon-sized Mobiloil Arctic Special (Light). $150–175

Socony Vacuum Oil Co., rectangular gallon-sized generic can with a place for a glued label. $25–30

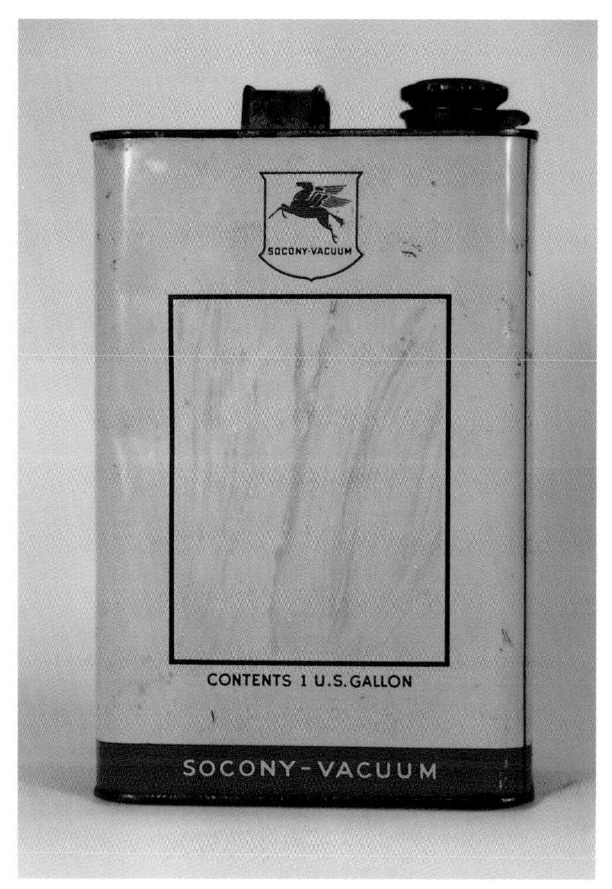

Socony Vacuum Oil Co., rectangular gallon-sized generic can with a place for a glued label. $25–30

Gargoyle, Socony Vacuum Oil Co., five-quart steel Mobiloil can. $105–125

Socony-Vacuum Oil Co. Inc., five-quart steel Mobiloil can. $85–100

Socony Mobil Oil Co. Inc., five-quart steel Mobiloil Special can. $100–125

Socony Mobil Oil Co. Inc., five-quart steel Mobiloil can $65–8

Socony Mobil Oil Co. Inc., five-quart steel Mobiloil can. $75–

Socony Mobil Oil Co. Inc., five-quart aluminum Mobiloil can. $50–65

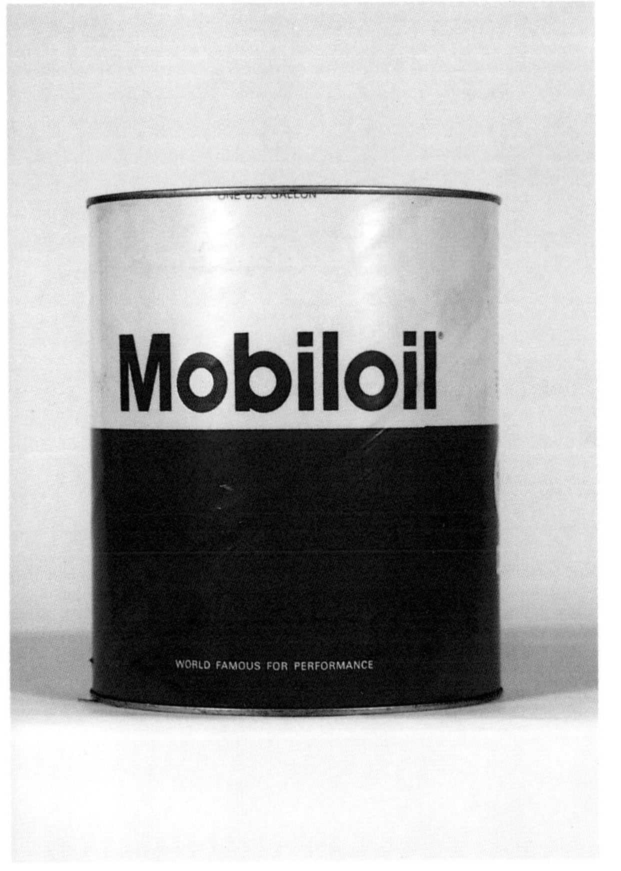

Mobil Oil Corp., composite gallon-sized Mobiloil can. $15–25

Socony Mobil Oil Co. Inc., five-quart steel Mobiloil Super can. $75–90

White Eagle Oil and Refining Co., three-gallon Keynoil pour can. $80

Wadhams, three-gallon Big "W" motor oil can. $75–100

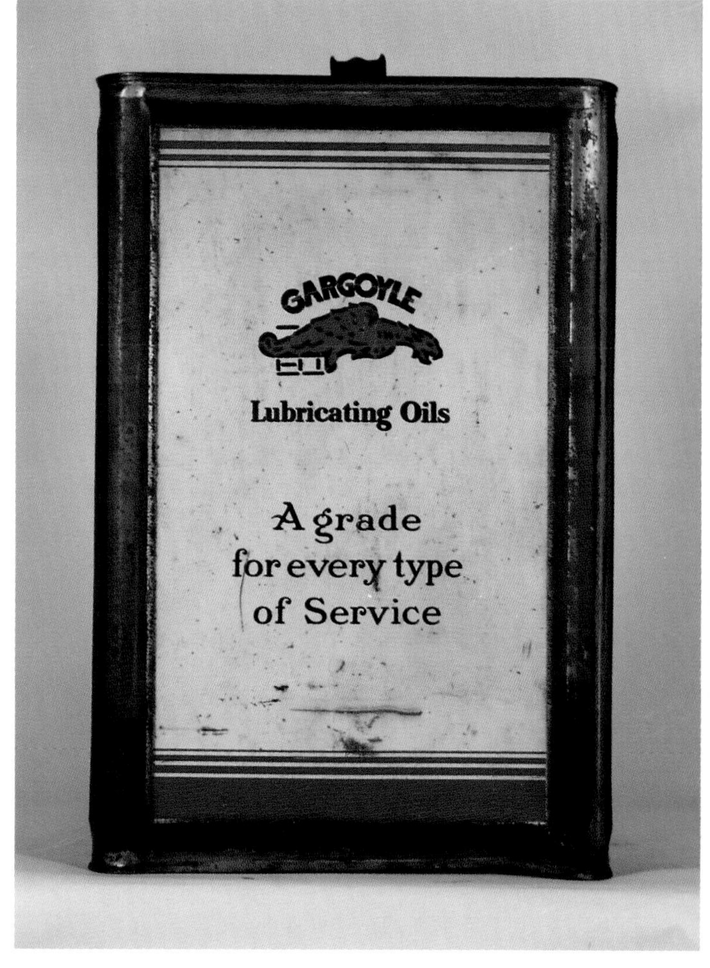

Gargoyle Vacuum Oil Co., five-gallon square Lubricating Oils can. $65–85

Socony-Vacuum Oil Co., five-gallon D.T.E. Oil Heavy Medium can. $35–50

Socony-Vacuum Oil Co. Inc., five-gallon Mobiloil can. $25–40

Socony Mobil Oil Co. Inc., five-gallon Lubrite Motor Oil can. $20–35

Socony Mobil Oil Co., five-gallon Mobiloil can. $25–35

II. Lube and Grease Cans

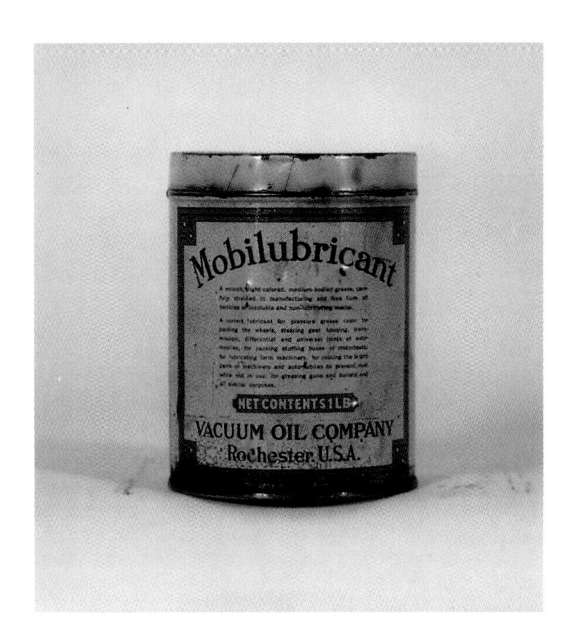

Vacuum Oil Co, Rochester U.S.A.,
1-lb. Mobilubricant. $100

Vacuum Oil Co., Inc., New York, U.S.A.,
1-lb. Mobilubricant. $100

Vacuum Oil Co, Rochester U.S.A., 1-lb. Mobilubricant Handy Package. $150

Magnolia Petroleum Co., Texas. 1-lb. Magnolia Axle Grease, embossed lid w/can. $250

Socony Vacuum Oil Co., Inc., with gargoyle. 1-lb. Mobilgrease can. $115–135

Socony Vacuum Oil Co., Inc. 1-lb. can Mobilgrease No. 4. $45–60

White Eagle Division of Socony Vacuum Oil Co., Inc.
1-lb. grease can. $50

Socony Vacuum Oil Co. 1-lb. can Mobilgrease #4. $30

Socony Mobil Oil Co., Inc. 1-lb. Mobilgrease can. $35–50

Socony Mobil Oil Co., Inc. 7-oz. Mobilgrease No. 8 can. $35–50

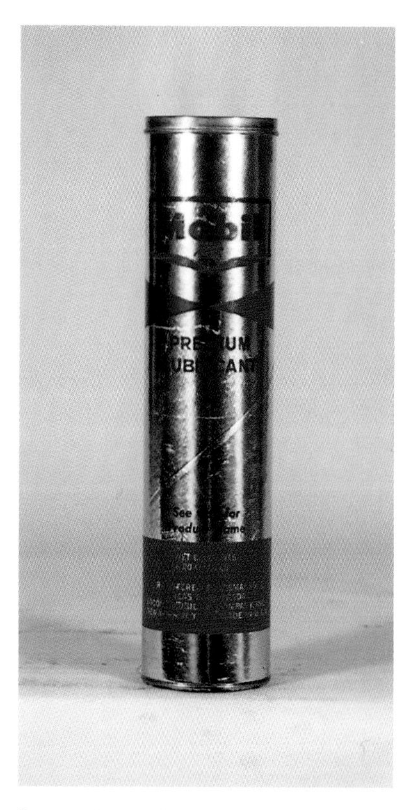

Socony Mobil Oil Co., Inc. 20-oz. tube Mobilgrease. $20–25

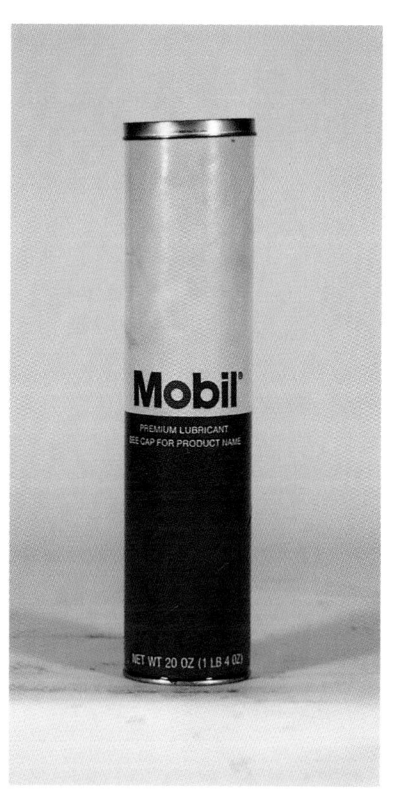

Mobil Oil Corp. 20-oz. tube Mobilgrease. $10–12

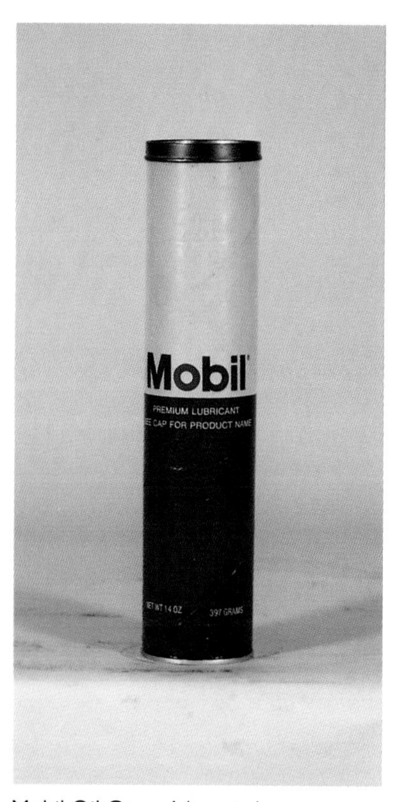

Mobil Oil Corp. 14-oz. tube Mobilgrease. $4–8

Gargoyle, Socony Vacuum Oil Co. 2-lb. Mobilgrease "FS" (for Ford and Lincoln Zephyr springs only) can. $150

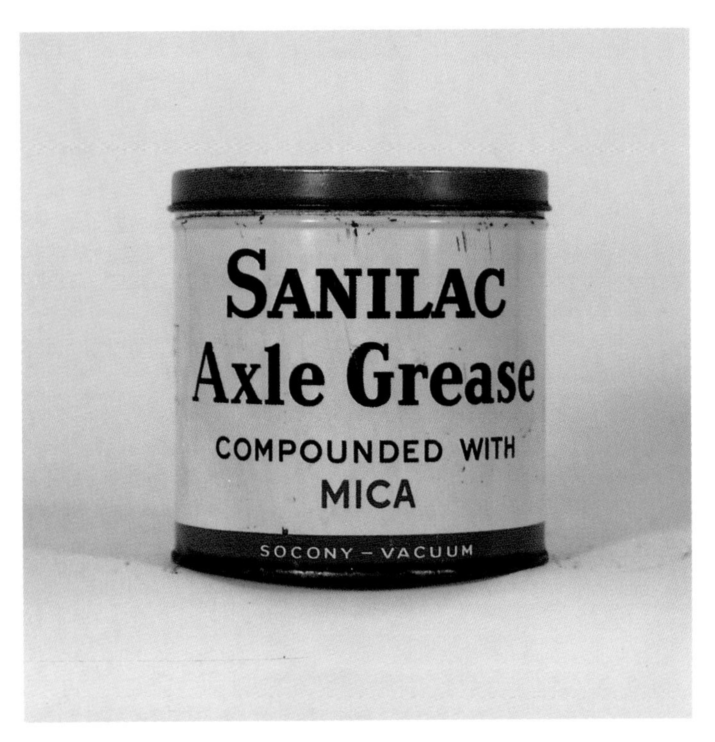

Socony-Vacuum. 3-lb. Sanilac Axle Grease compounded with MICA. $100

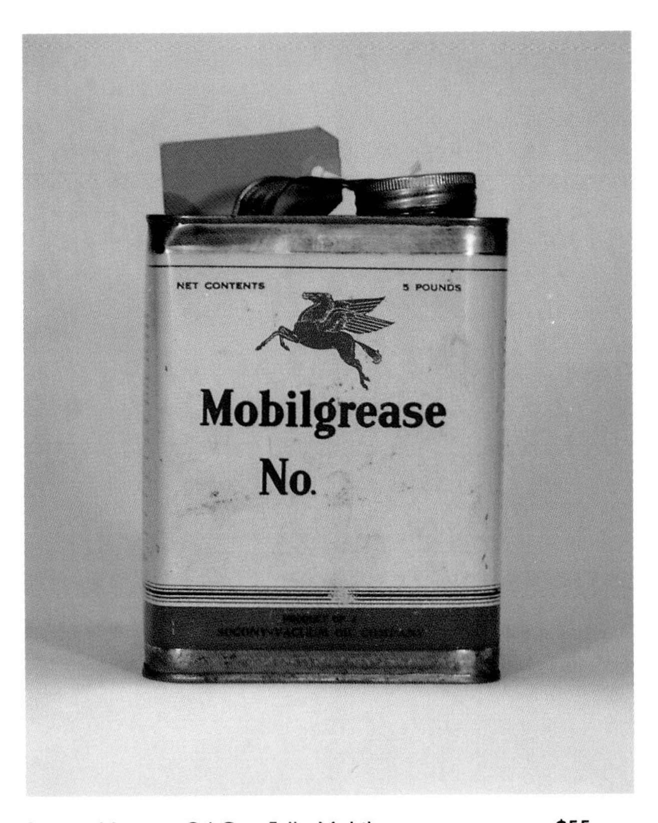

Socony-Vacuum Oil Co., 5-lb. Mobilgrease square can. $55

Gargoyle, Vacuum Oil Co., 5-lb. Mobilgrease square can. Notice the stamp signifying Socony-Vacuum Oil Co. Inc., placed below the Gargoyle—this can is from 1929 to 1930, just before the merger. $100

Socony-Vacuum Oil Co., Inc. 5-lb Mobilgrease Aero square can. $45–60

Socony-Vacuum Oil Co., Inc. 5-lb. Mobilcote 270 square can. $45–60

Wadhams Oil Co. 5-lb. Wadhams grease can. $45–50

Socony-Vacuum Oil Co., Inc. 5-lb. Mobilgrease can. $25–35

White Eagle division of Socony-Vacuum Oil Co., Inc. 5-lb. White Eagle Cup grease can. $60

Socony Mobil Oil Co., Inc. 5-lb. Mobilgrease can. $15–25

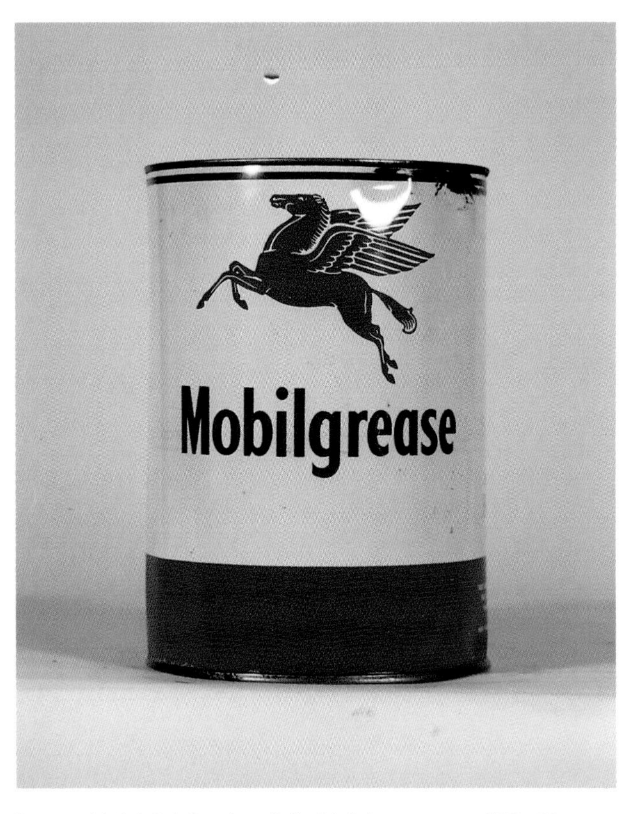

Socony Mobil Oil Co., Inc. 5-lb. Mobilgrease can. $10–20

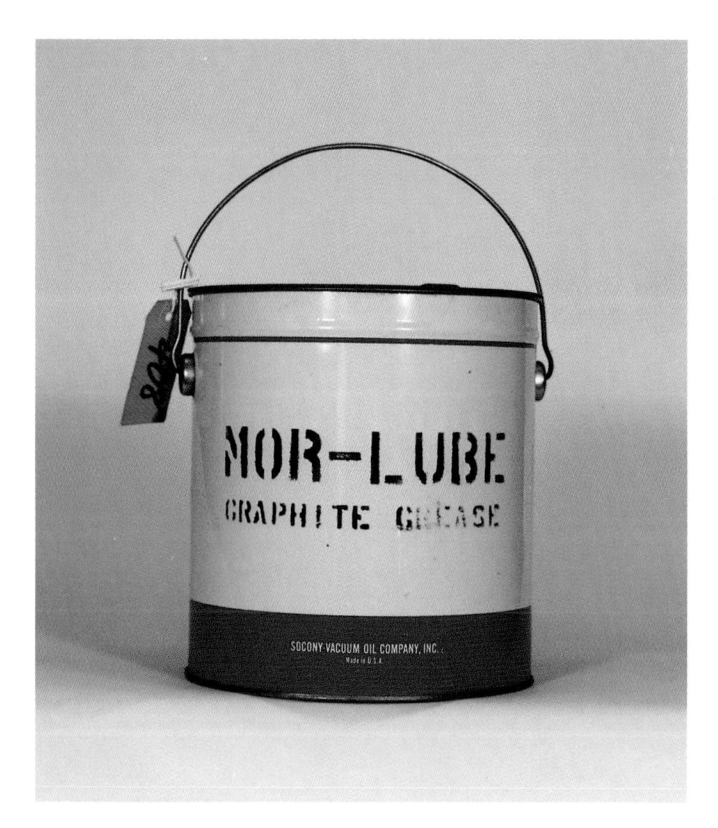

Socony-Vacuum Oil. Co., Inc. 10-lb. MOR-LUBE graphite grease. $65–90

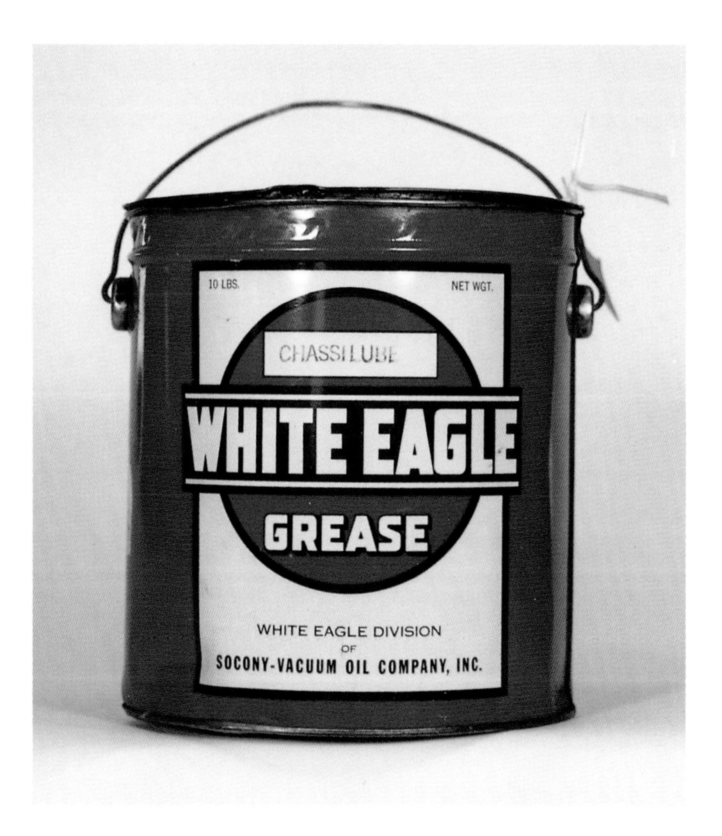

White Eagle Division, Socony-Vacuum Oil Co., Inc. 10-lb. Chassilube White Eagle grease can. $35

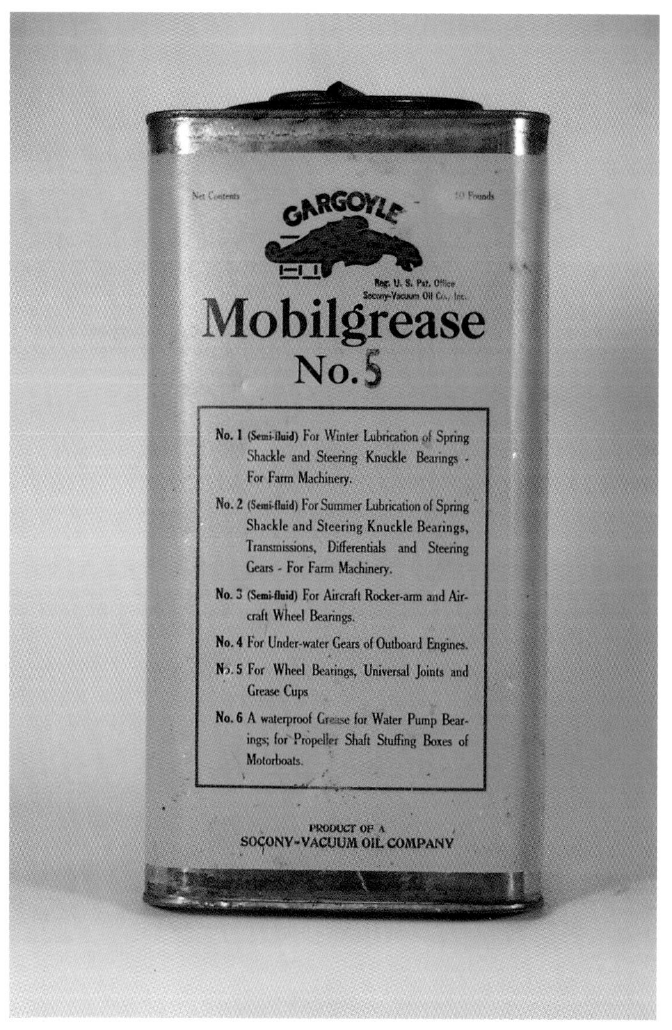

Socony-Vacuum Oil Co. 10-lb. Mobilgrease No. 5 can. $200

White Eagle Division, Socony-Vacuum Oil Co., Inc. 25-lb. Chassilube White Eagle grease can. $50

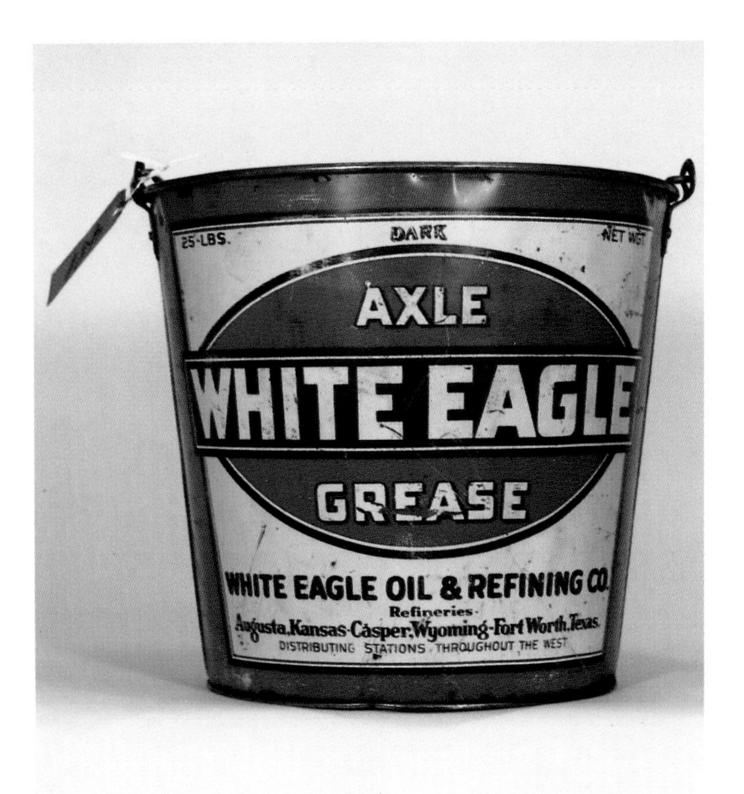

White Eagle Oil & Refining Co., 25-lb. White Eagle Axle grease pail. $250–300

Socony Mobil Oil Co., 38-lb. Mobilgrease can. $20–30

Socony Mobil Oil Co., 38-lb. Mobilgrease can. $20–30

Boston Gear Works Inc., for Socony Vacuum. Cylinder Oil 600w can with paper label. $150

General Petroleum Corp., for McCulloch Motors Corp., one-pint Gear Case Oil Mobilube "C" SAW 140. $25

Socony-Vacuum Oil Co., Inc. one-quart Military Spec. lube for Aircraft & Machine Guns, dated May 1943. $45–60

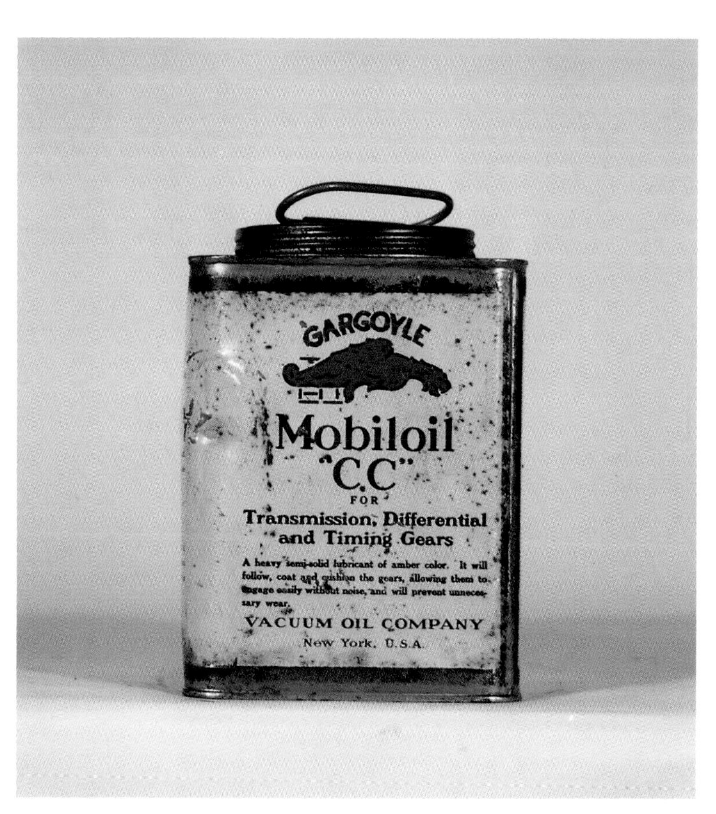

Gargoyle, Vacuum Oil Co., 5-lb. Mobiloil "CC," square can. $100–125

Bottom of 100-lb can with date.

Socony-Vacuum Oil Co., front view of 100-lb. can Mobil Oil "C". $100–125

Top of can.

Standard Oil Co. of New York, 1-gallon Socony Gear Oil can. $100

Gargoyle, Socony-Vacuum Oil Co., Inc. 100-lb Mobilgrease No. 2 can. Has Bung at bottom for valve. $100–125

Socony-Vacuum Oil Co., Inc. 120-lb. Mobilube can. $65–85

Socony Mobil Oil Co., Inc. 120-lb. Mobilube can. $50–75

Gargoyle, Socony-Vacuum Oil Co., Inc. 2-lb. Mobilgrease UW (a waterproof lubricant for underwater gears) with a screw-type dispenser. $100–125

Socony-Vacuum Oil Co., Inc. 2-lb. Mobilgrease UW with a screw-type dispenser. $75–90

Gargoyle, Socony-Vacuum Oil Co., Inc. 2-lb. Mobilube GX 90, Outboard Gear Oil can. $35–45

Socony-Vacuum Oil Co., Inc. 2-lb. Mobilube GX 90, Outboard Gear Oil can. $30–40

Socony Mobil Oil Co., Inc. 2-lb. Mobilube for gears can. $25–35

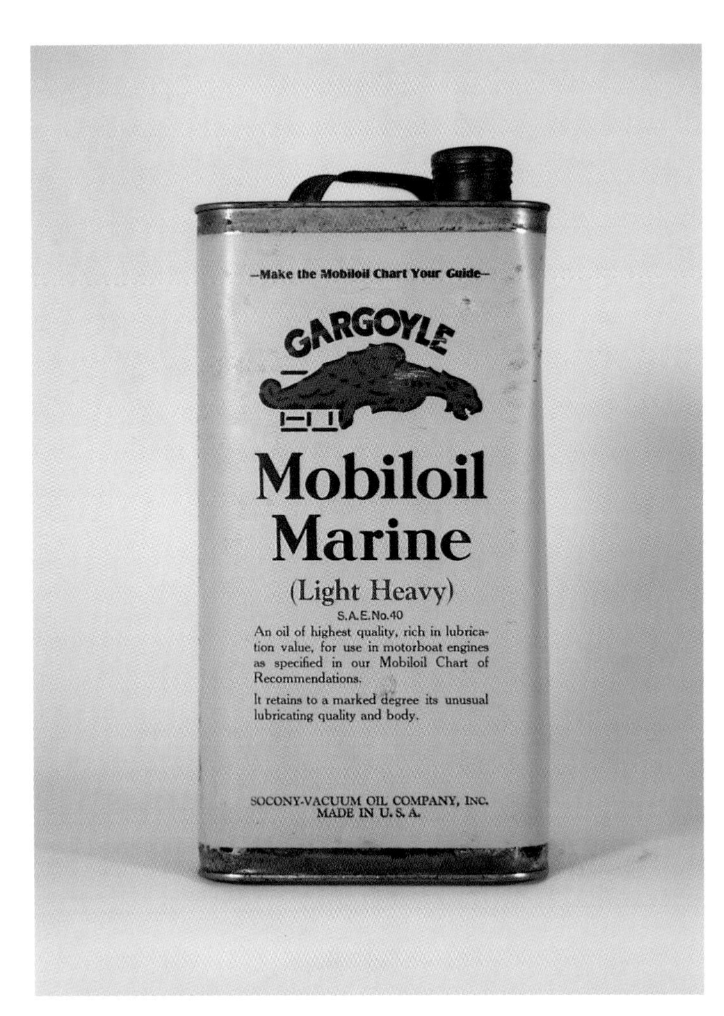

Gargoyle, Socony-Vacuum Oil Co., Inc. one-gallon. Mobiloil Marine (Light Heavy) SAE No. 40 can. $250

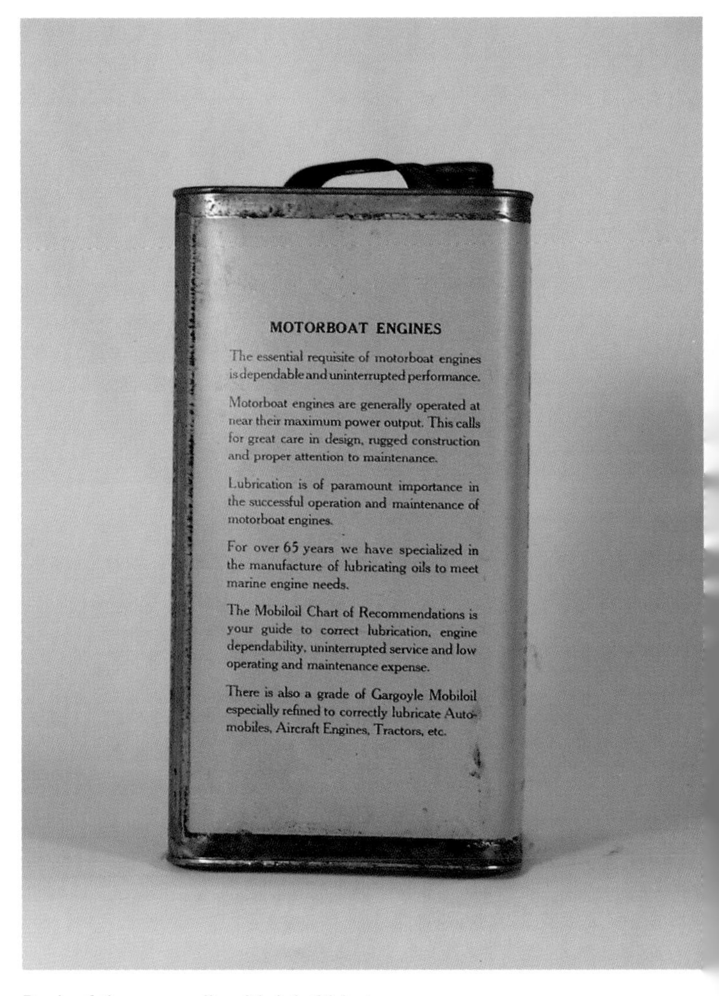

Back of the one-gallon Mobiloil Marine can.

Socony-Vacuum Oil Co., Inc., one-quart Mobiloil Outboard can. $35–50

Socony-Vacuum Oil Co., Inc., one-quart Mobiloil outboard can. $30–40

Socony Mobil Oil Co., Inc., one-quart Mobiloil outboard can. $30–40

Socony Mobil Oil of Canada Ltd. imperial one-quart Mobiloil outboard can. $35–50

Socony Mobil Oil Co., Inc., one-quart Mobiloil outboard can. $30–45

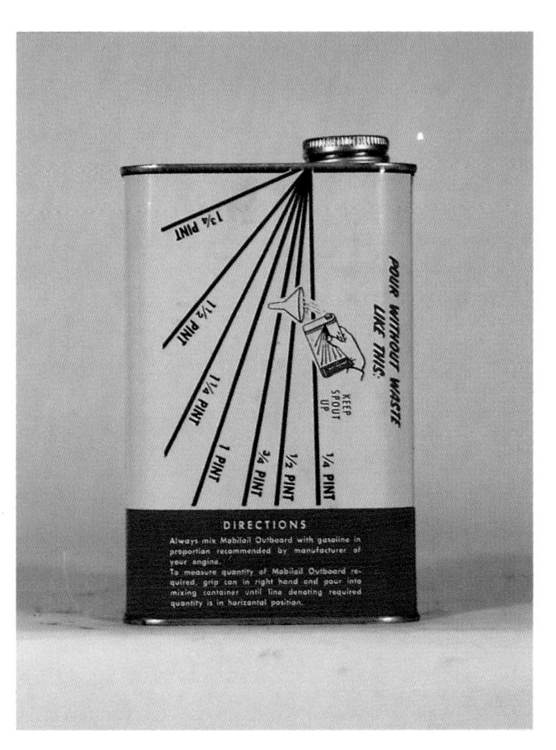

Back of the Socony Mobil outboard oil can. Has diagram for mixing with gasoline.

Socony Mobil Oil Co., Inc., one-quart Mobiloil Outboard can. $25–35

Mobil Oil Corp., one-pint Mobil outboard can. $10–15

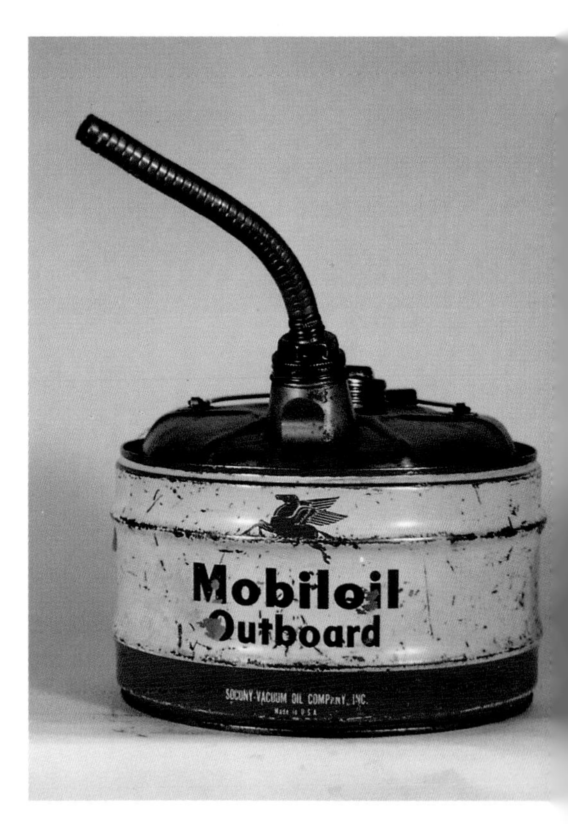

Socony-Vacuum Oil Co., Inc., two-gallon Mobiloil outboard gas can. $65–80

III. Signs, Racks and Dispensers

35 1/2" x 60" porcelain sign. $160

Octagonal White Eagle porcelain sign. $200

20" x 28" oval porcelain Keynoil sign. $800

White Star paint sign from the 1920s. $200

3' x 5' Mobiloil porcelain sign. $1200

3' porcelain embossed pegasus. $1500

One-sided porcelain badge, used when Socony-Vacuum bought out Wadhams. This sign was attached to the base of the Wadhams sign. 17 1/4" x 18 3/4". $225–275

Lube requirement plate off of a Cultitractor recommending Gargoyle Oil. $40–55

Two-sided painted sign. 11"x 20". $150–175

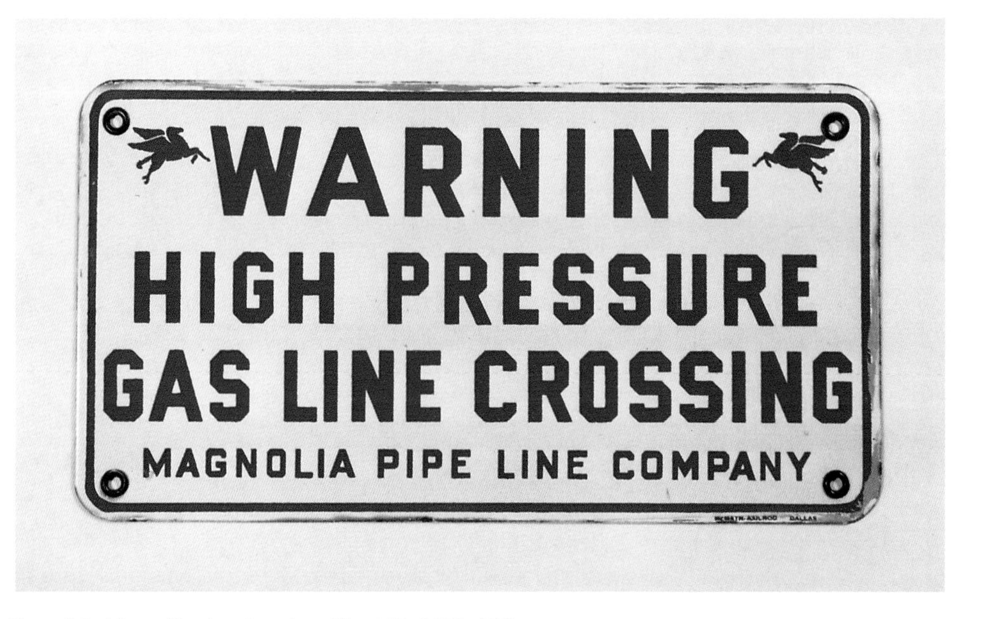

Porcelain Magnolia pipe line sign. 8" x 15". $125–150

Painted Mobilgas "No Smoking" sign. 5" x 20". $100–125

Set of wooden letters for the outside of a station. Late 1950s to early 1960s (has the original crate). $400–500

One-sided porcelain Mobilgas rest room sign. 7 3/4" x 7 3/4" $300–350

Porcelain Mobilgas Special pump plate, i.r. 1947. 12 1/4" x 12 1/4". $125–150

Porcelain Mobiloil plate, i.r. 1946. 12 1/4" x 12 1/4". $300–325

Porcelain Mobilfuel Diesel, i.r. 1951. 12 1/4" x 12 1/4". $300–325

Porcelain Mobil Regular pump plate. 12" x 13 3/4". $100–115

Mobilgas Special enameled pump plate, 7" x 11 1/4". $145–165

Porcelain Mobil Premium pump plate. 12" x 13 3/4". $115–130

Free-standing oil bottle rack. $900–1000, without the bottles.

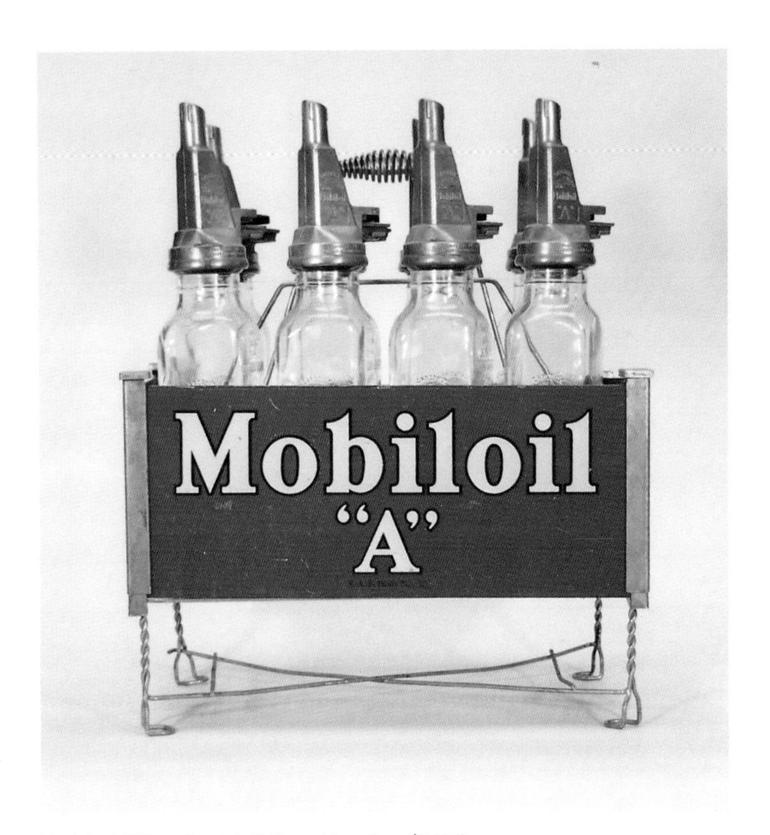

Mobiloil "A" rack with 8 filpruf bottles. $1850

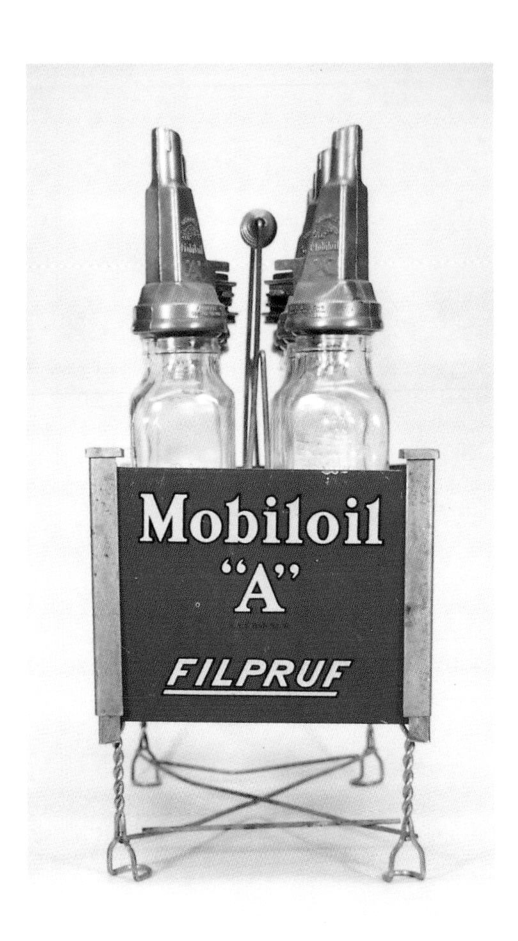

Side view of Mobiloil "A" rack.

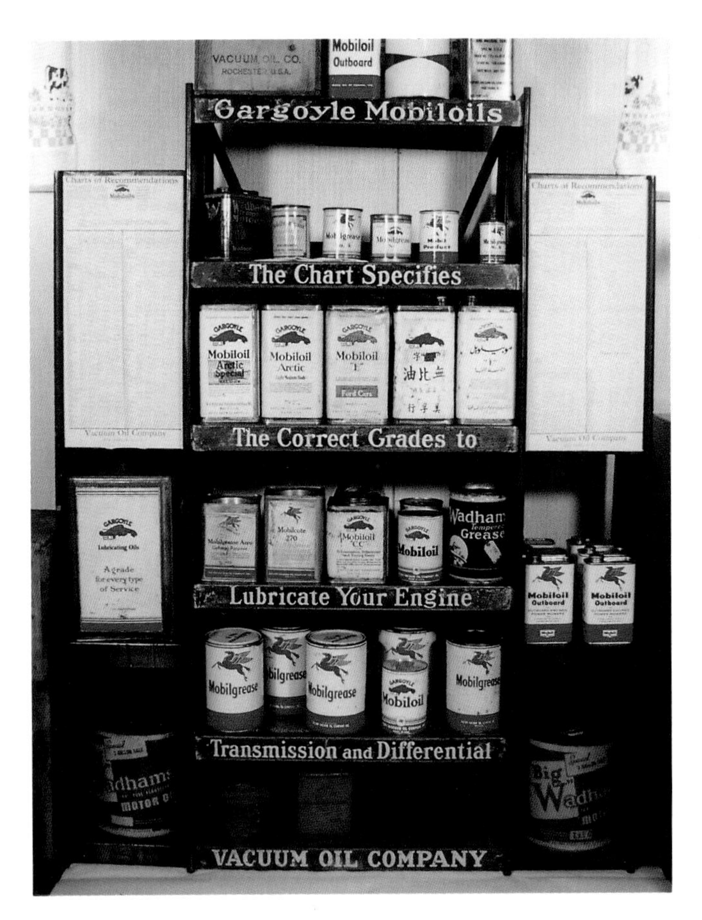

Vacuum Oil Co., Gargoyle store display rack w/ reproduction charts, rack is from the 1920s. $1200–1400, rack only.

Side view of the store display rack.

Socony Vacuum Oil Co., Inc. Painted rack topper. 16 1/2" diameter. $125–150

36" x 74 1/2" Red Horse Shop wooden display rack. $250–275

Socony Vacuum Oil Co., Inc., Tavern brand small tin shelf for Tavern candles. $75–90

Socony Vacuum Oil Co., Inc. 2-sided wooden Tavern shop sign. $140

Socony-Vacuum Co., 1940s Telecron clock with a 14" diameter face. $425–450

1-piece Gargoyle globe. $1800

Mobilgas Ethyl globe, 2 lenses with steel body; 16 1/2" diameter. $750

Vacuum Oil Co., enamel Mobilgrease lubster sign, 7" diameter. $165–200

Restored Gilbert Barker Oil Lubster with reproduction decal and Original 10" porcelain sign. Lubster alone, $150. 10" porcelain sign alone, $165–225

Restored oil lubster with reproduction sign. The top of the lubster has a stamped Gargoyle drain cover, and on the back corner of the lubster embossed are the words "PROPERTY OF VACUUM OIL CO." $225–250

Stamped Gargoyle drain cover.

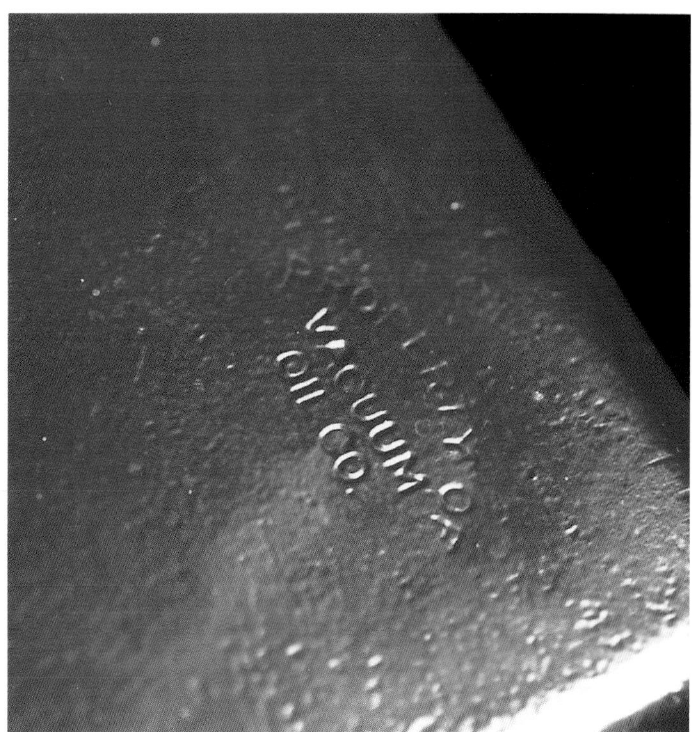

The embossed wording "PROPERTY OF VACUUM OIL CO."

Restored Martin & Schwartze Model 80 with reproduction sign and globe. $800–1000

Portable lubster that fits on a 120-lb. barrel, with an original Socony Mobil Oil Co., Inc. barrel. $75–100

Restored Martin & Schwartze Model 80 with original script top and reproduction sign. $1000–1200

One-quart Mobil Filpruf diamond-shaped bottle with embossed gargoyle. $115–130

One-quart oil bottle Mobiloil "A" SAE 10. $50–65

One-quart Marquette oil bottle Mobiloil "A" SAE 30. $50–65

Embossed oil spouts. Left, Wadhams tempered motor oil. $15–20. Right, Gargoyle Mobiloil Arctic, $15–20.

Half-pint Mobiloil pour can. $55

Vacuum Oil Co., barrel valve shut-off spout with a Gargoyle handle. $175

Gargoyle handle for the barrel valve shut-off spout.

Socony-Vacuum Oil Co., Inc. Wooden box for six one-gallon Mobiloil Arctic Special cans. $55–75

Gargoyle, Vacuum Oil Co., New York U.S.A. Wooden box for ten one-gallon Mobiloil "C" cans. $75–100

Gargoyle, Vacuum Oil Co., Rochester U.S.A. Wooden box for one five-gallon Mobiloil "A" can. $75–100

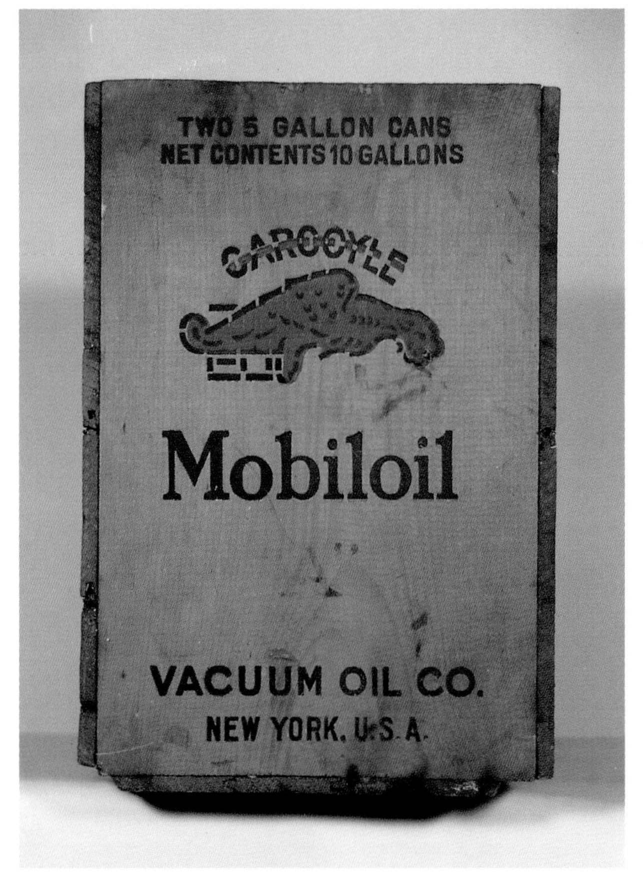

Gargoyle, Vacuum Oil Co., New York U.S.A. Wooden box for two five-gallon Mobiloil cans. $75–100

Certified five-gallon tank wagon can. $40–65

Wadhams embossed certified ten-gallon tank wagon can. $50–75

Standard Oil of New York., embossed tank wagon pour can.

8' Pegasus "Cookie Cutter" sign dated IR 55, facing to the right. $2500–2800

IV. Automotive Repairs and Care

Socony-Vacuum Oil Co., Mobilgas radiator "Winterfront". $100–125

Socony Mobil Oil Co., Inc., one-quart glass jar of Mobil Radiator Flush. $35–45

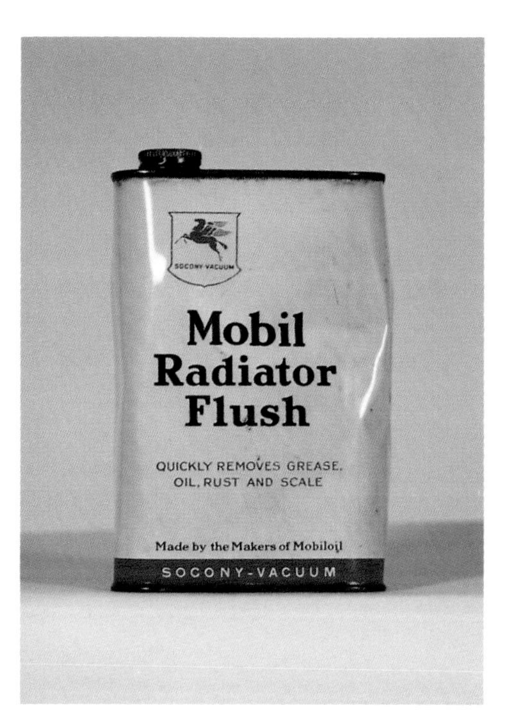

Socony-Vacuum Oil Co., one-quart Mobil Radiator Flush. $35–55

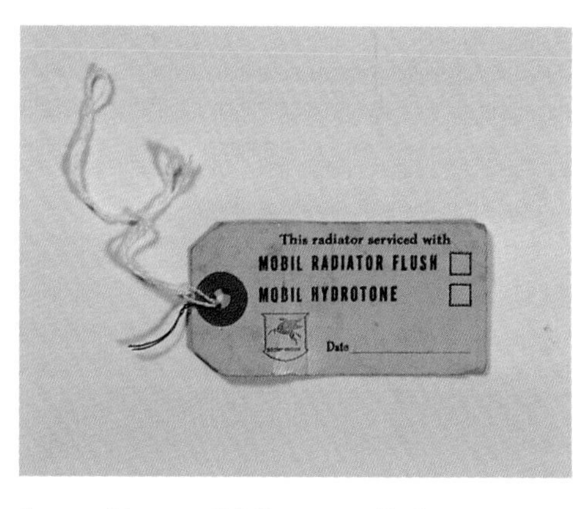

Socony-Vacuum Oil Co., paper Radiator service tag. $6–15

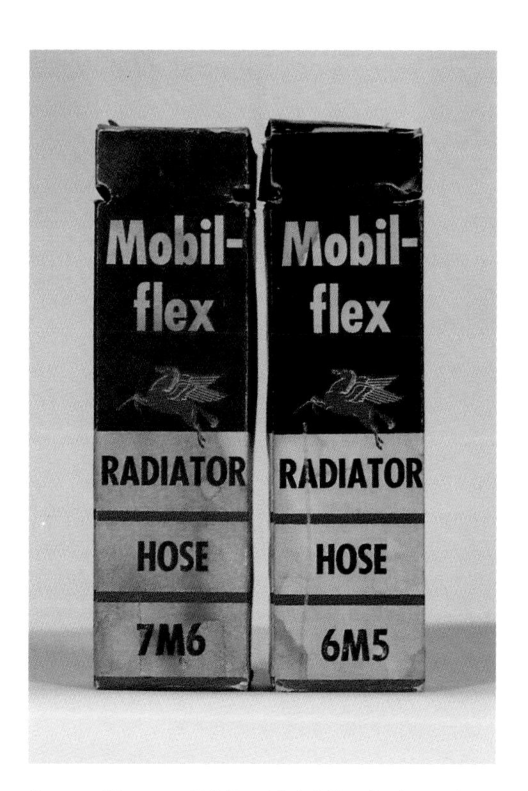

Socony-Vacuum Oil Co. Mobil-flex Radiator hoses with boxes. $25–40 each.

8-oz Mobil Stop Leak cans. Left, Socony Vacuum, $45–55. Right, Socony Mobil, $35–45.

8-oz Mobil Hydrotone cans. Left, Socony-Vacuum, $45–60. Right, Socony Mobil, $35–50.

Socony-Vacuum, one-gallon Mobil Freezone pour can. $75–100

Socony-Vacuum, one-quart Mobil Freezone pour can. $50–65

Socony Mobil, one-quart steel Freezone can. $35–50

Socony Mobil, one-quart steel Permazone can. $25–40

Mobil Oil Corp, one-quart steel Permazone can. $15–25

Socony Mobil, one-quart steel Freezone can. $25–40

Socony Mobil, one-gallon steel Permazone can. $30–45

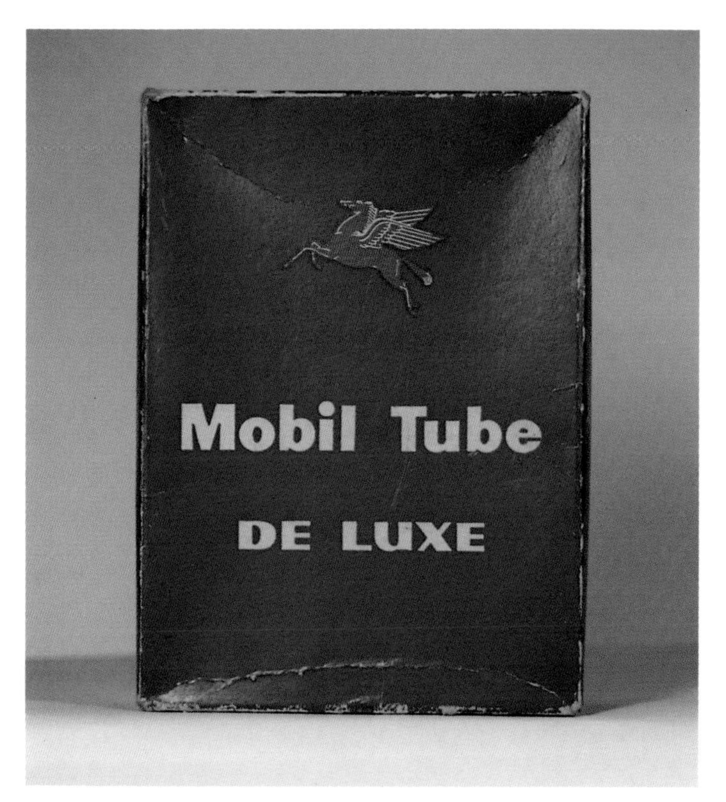

Socony-Vacuum, Mobil Tube (inner tube) box. $25–40

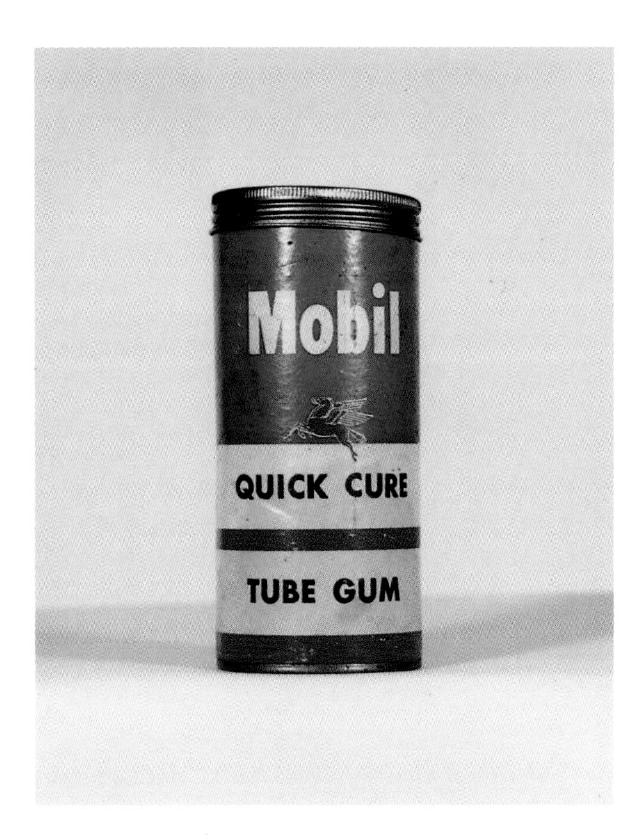

Socony-Vacuum. Service station size tube repair kit, cardboard can with screw-on lid. $85–100

Socony-Vacuum. Service station size tube repair kit, steel can with push-on lid. $50–75

Mobil Tread depth Safe-T-Scope. $85–110

Mobil Corp., tire stand. $50–75

N.O.S. Mobil Tire Stand. $110–135

Mobil Tire Display "Center". $65–85

Socony Specialities, Inc. Pre-1931, 4-oz Socony Upperlub oil can. $50–75

Side view of Socony Upperlub can.

Socony Mobil, one-pint Hydraulic Brake fluid can. $50

Socony Mobil, 4-oz Mobil Upperlube can. $30–45

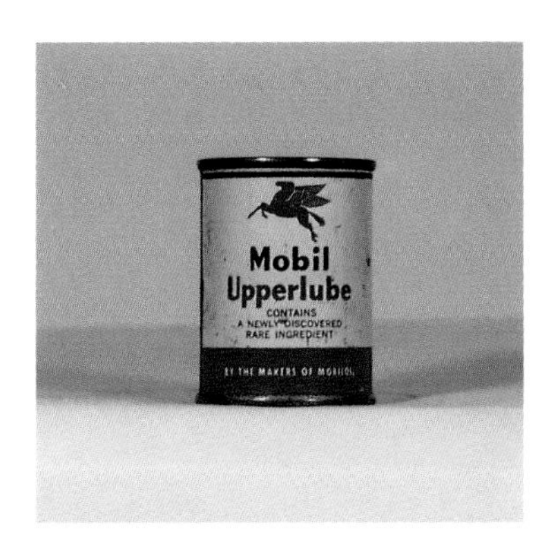

Socony-Vacuum, 4-oz Mobil Upperlube can. $35–50

Socony Mobil, 4-oz Upperlube can. $25–40

Upperlube can opener with patent number 2184830. $50

Mobil Oil Corp. Left, 4-oz. Upperlube can, $15–30. Right, 8-oz Upperlube super, $15–30

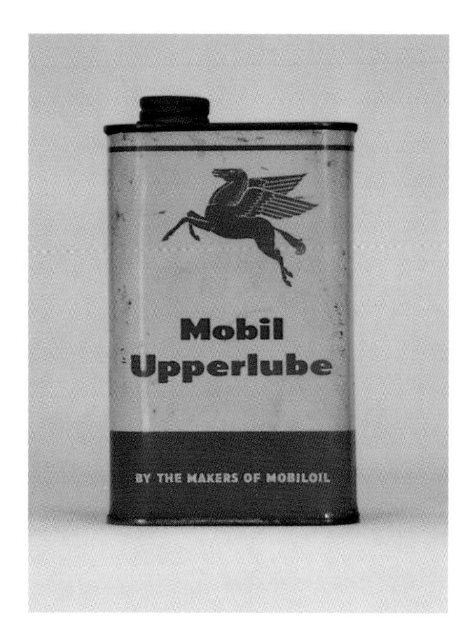

Socony Mobil. One-pint Mobil Upperlube can. $45

Socony Mobil. One-pint Mobil Upperlube can. $25–40

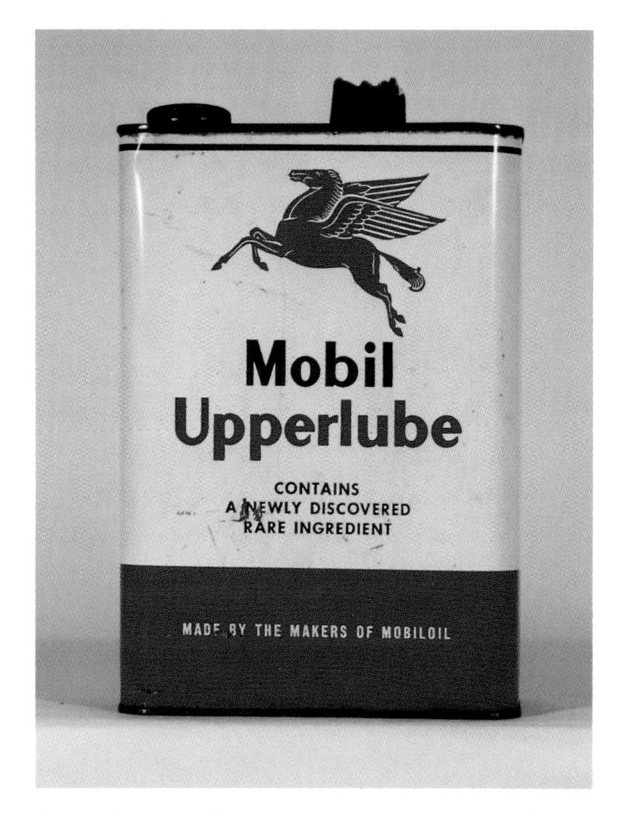

Socony-Vacuum. One-gallon Mobil Upperlube can. $25–40

Socony Mobil. One-gallon Mobil Upperlube can. $20–35

Mobil Start-O-Scope battery tester. $150–175

Side view of Start-O-Scope.

Socony-Vacuum. Mobil Lustre cloth can. $35–50

Socony-Vacuum. Tavern Brand Lustre cloth can. $40–55

Socony-Vacuum. Mobil Lustre cloth can. Note the off-white color of the can. $25–40

Mobil Lustre Cloth cans. Left, Socony-Vacuum, $25-35. Right, Socony Mobil, $20–30

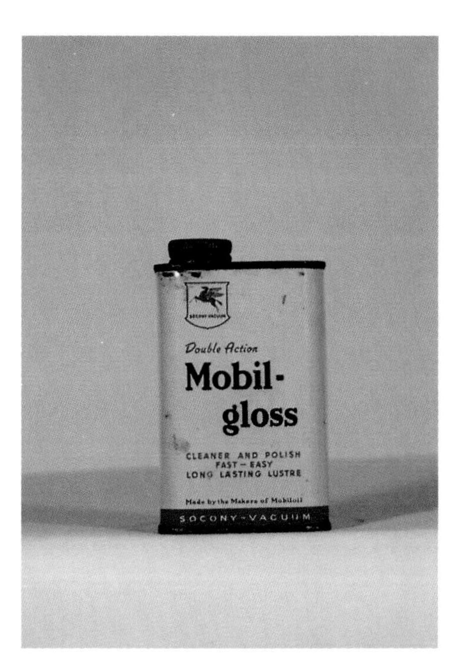

Socony-Vacuum. 8-oz Mobil-gloss can. $40–50

Socony-Vacuum. 10-oz Mobilgloss can. $35–50

Socony Mobil. One-pint Mobilgloss can. $25–40

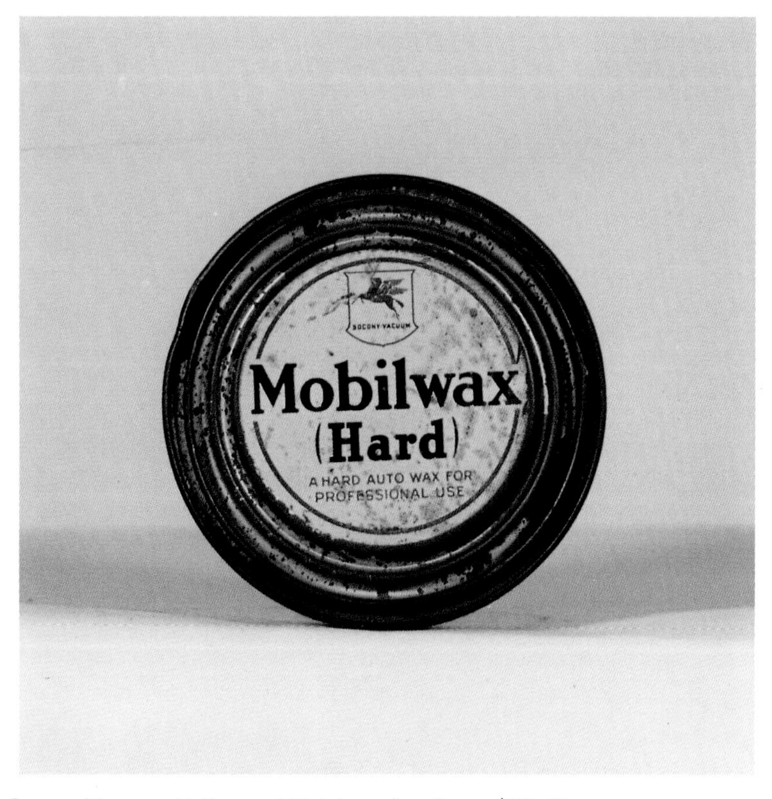

Socony-Vacuum. Half-pound Mobilwax (hard) can. $30–45

Socony-Vacuum Oil Co., Mobil Window Spray can. $15–20

Socony-Vacuum. One-pint Sovasuds Car Wash jar. $40–55

Socony-Vacuum. All purpose cleaner. $35–50

Socony Mobil. One-pint Sovasuds Car Wash jar. $35–50

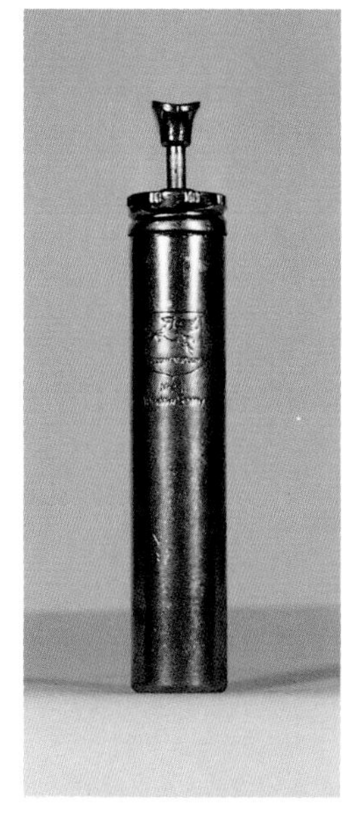

Socony-Vacuum. Service station attendant's window sprayer with embossed pegasus. $90–115

Left, Mobil Corp. windshield cleaner spray bottle, $18–25. Right, Socony-Vacuum Co. window spray bottle, $125.

Flying Horse summer weave baseball-style giveaway hat. $25–35

Left, Mobil summer hat, $35–45. Right, Mobil winter hat with flaps, $45–55

Mobilgas hard brimmed attendant's hat with winter frame. $145–175

Mobil 8-point attendant's hat with summer open weave frame. $165–185

Mobil 8-point attendant's hat with winter frame. $165–185

Mobil credit card clipboard. $25–35

Socony-Vacuum. One-pint can Bug-a-boo Super Insect Spray. $35–50

Socony-Vacuum. One-quart Bug-a-boo Super Insect Spray can. $35–50

Socony-Vacuum. Small Bug-a-boo Insect Sprayer. $115–140

Socony-Vacuum. One-pint Bug-a-boo Insect Spray jar. Note: 3% D.D.T. $45–60

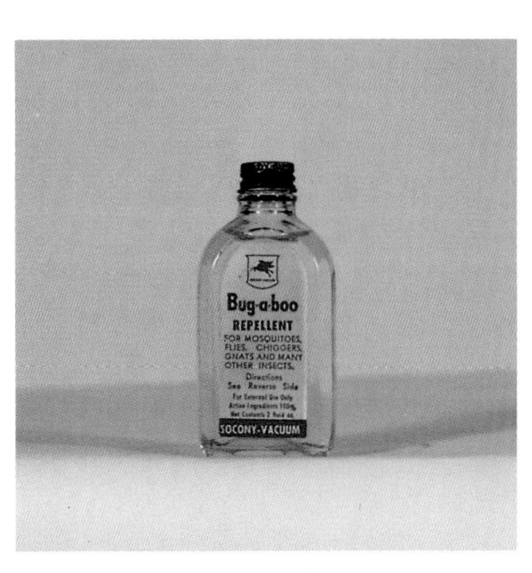

Socony-Vacuum. 2 oz Bug-a-boo Repellent. Note: 100% active ingredients. $45–60

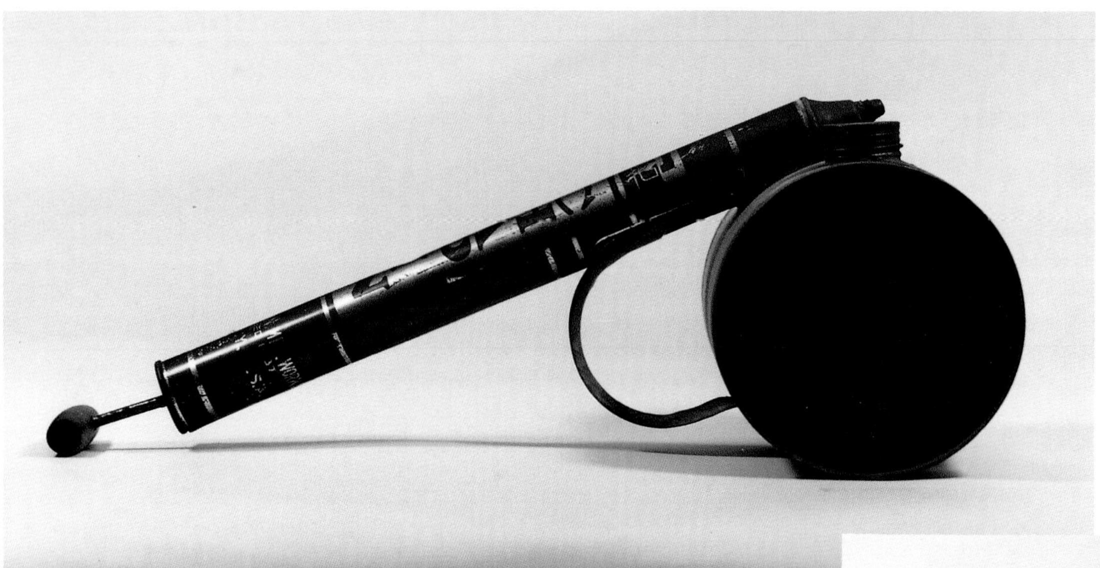

Socony-Vacuum. Large Bug-a-boo continuous sprayer. $75–100

Side view of continuous sprayer.

A nice display by George Deel. Far right: Bug-a-boo, Victory Garden Spray, $45.
Six bug-a-boo pin backs with the various stages of the bug's death. $35–60

Magnolia Petroleum Co. Magnolene Neatfoot Harness Oil container. $65

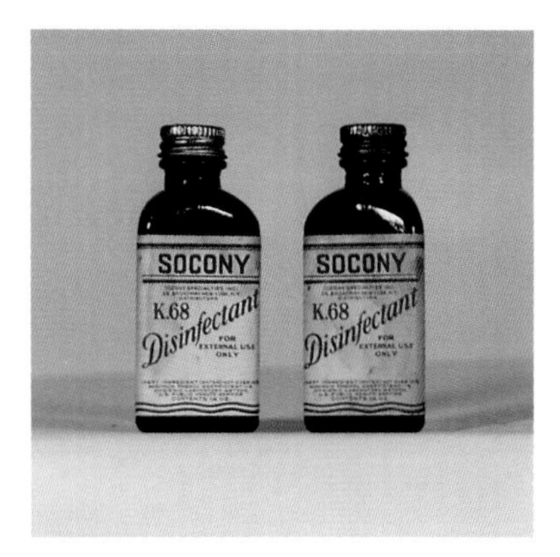

Socony. K68 disinfectant, 3-sided brown 1 1/4-oz bottle. $45–60 each.

Socony-Vacuum Oil Co. One-quart Sanilac Hand Separator Oil jar. $65

Socony Mobil. 4-oz. hand lotion jar. $30–45

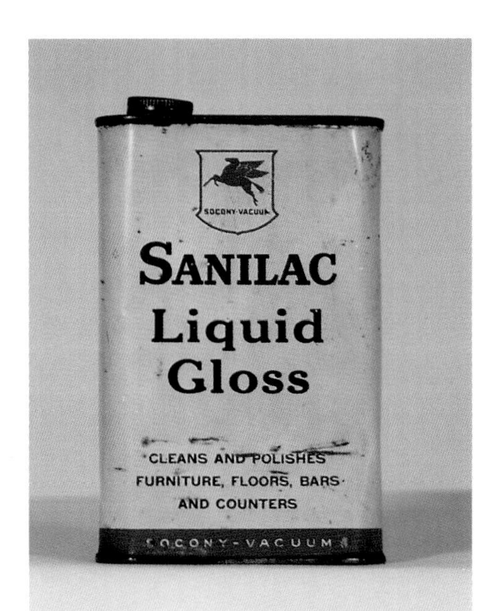

Socony-Vacuum. One-quart Sanilac liquid gloss can. $25–40

Socony Mobil. One-pint Mobil liquid burner cleaner can. $30–45

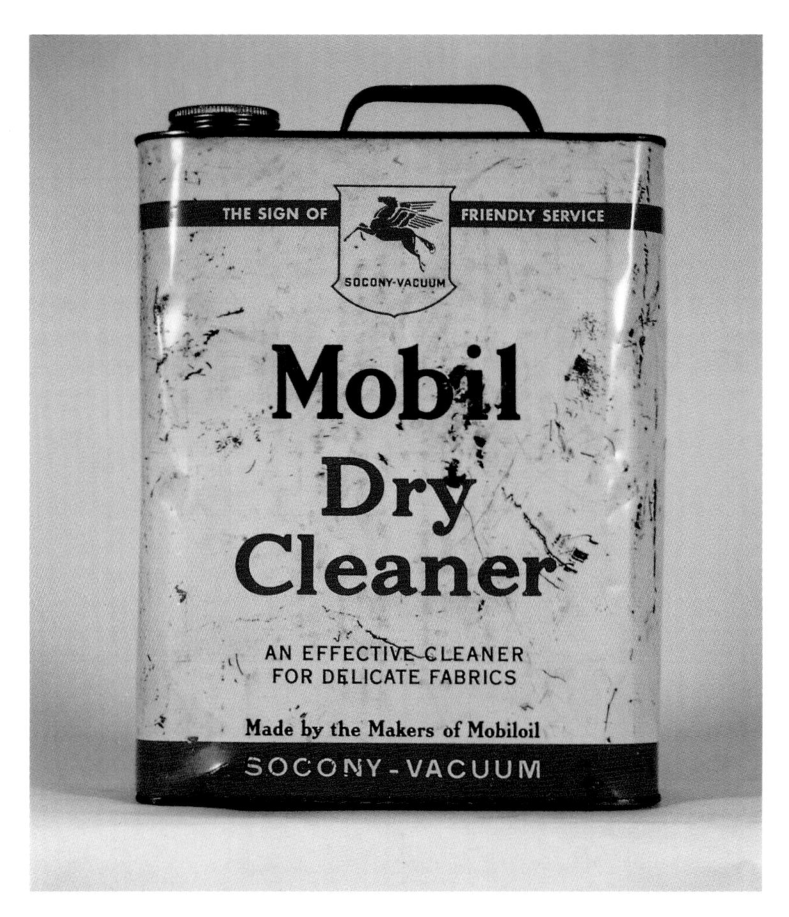

Socony-Vacuum. Two-gallon Mobil Dry Cleaner can. $30–45

Socony Mobil. One-pint Mobil oil burner cleaner can. $25–40

Socony Mobil. Left, 4-oz. Mobil Handy Oil can, $15–25. Right, Mobil Penetrating Oil can. $15–25

Socony-Vacuum. 4-oz. Mobil Handy Oil, rectangular can with steel top. $25–35

Socony Mobil. Left, 4–oz. Mobil Handy Oil can, $15–20. Right, Mobil Penetrating Oil can, $15–20.

Socony-Vacuum. 4-oz. Mobil Handy Oil, oval cans with steel tops. Left, $35–50. Right, $30–45.

Socony Mobil. Left, 4-oz. Mobil Handy Oil can, $15–25. Right, Mobil Penetrating Oil can, $15–25.

Socony Mobil. Left, 4-oz. Mobil Handy Oil can, $15–20. Right, Mobil Penetrating Oil can, $15–20.

Standard Vacuum. One-pint Mobil penetrating jar with a paper label. $40

Socony-Vacuum. One-pint Mobil penetrating jar with a paper label. $30

Socony Mobil. Left, 4-oz. Mobil Handy Oil can, $10–20. Right, Mobil Penetrating Oil can, $10–20.

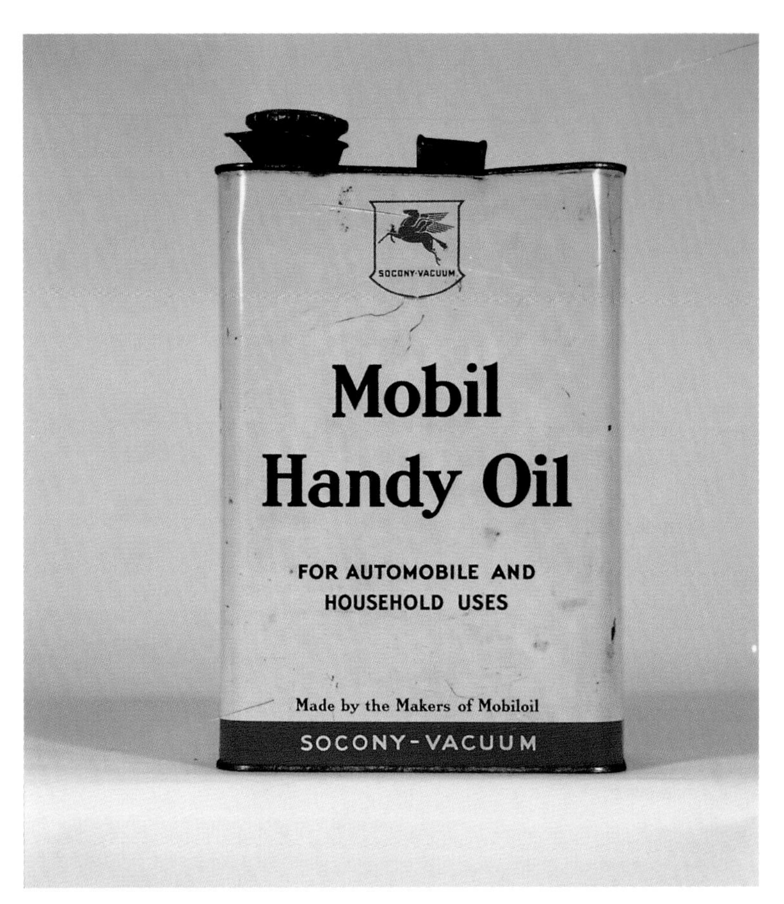

Socony-Vacuum. One-gallon Mobil Handy Oil can. $60–75

Socony-Vacuum. Gallon-sized Mobil Penetrating Oil. $25–40

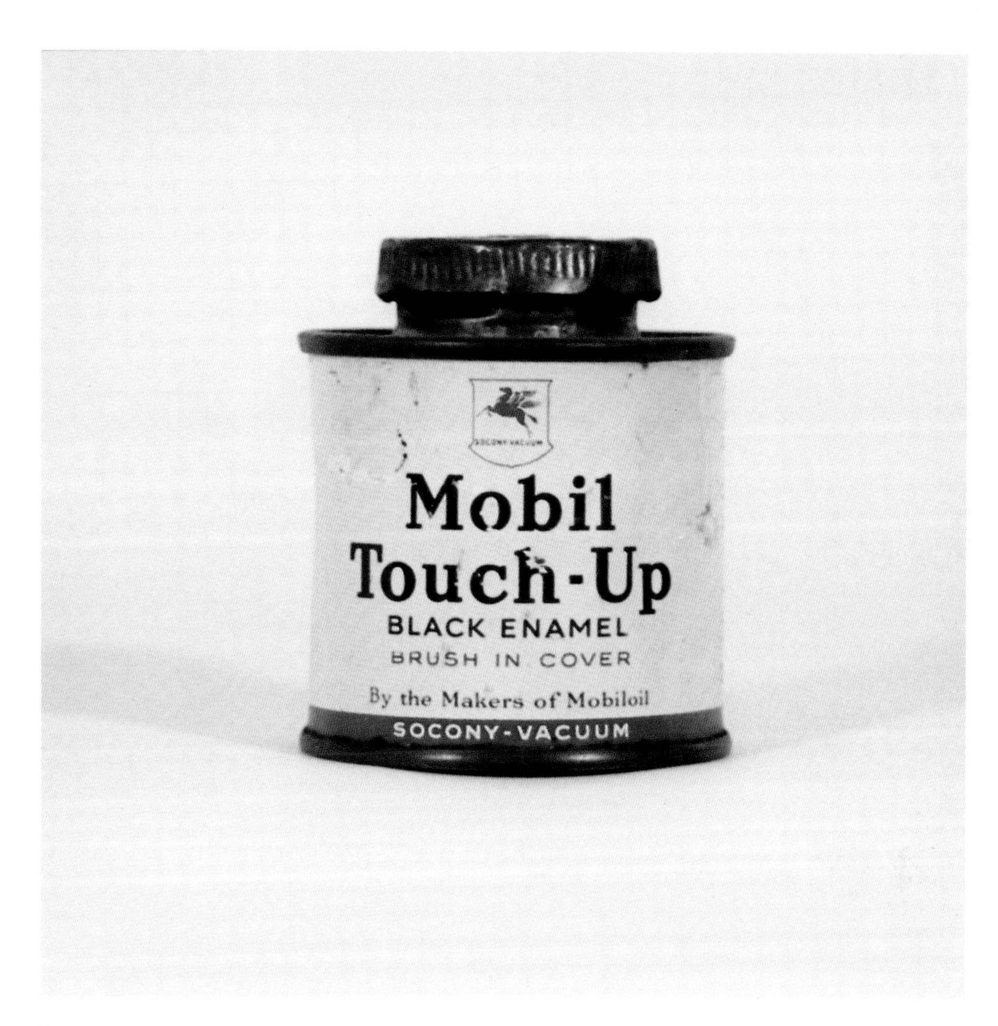

Socony-Vacuum. 5-oz. Mobil Touch-up black enamel paint. $100

Socony-Vacuum. 3-oz. Tavern Stain remover, oval can. $40–55

Socony-Vacuum. 4-oz. Tavern paint cleaner can. $50–75

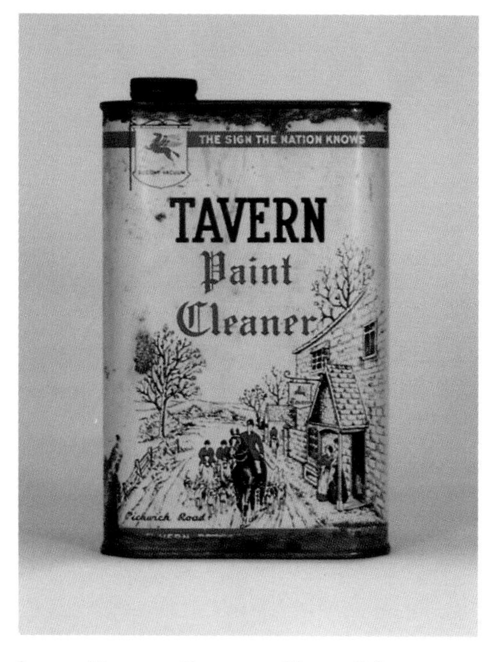

Socony-Vacuum. One-quart Tavern Paint Cleaner can. $30

Socony-Vacuum. 2 1/4-oz. tube with box. Tavern Electric Motor Oil. $35–50

Socony-Vacuum. One-pound Tavern Paraseal wax, in four quarter-pound slabs. $15–20

Socony-Vacuum. 8-oz. Tavern Electric Motor Oil can. $40–55

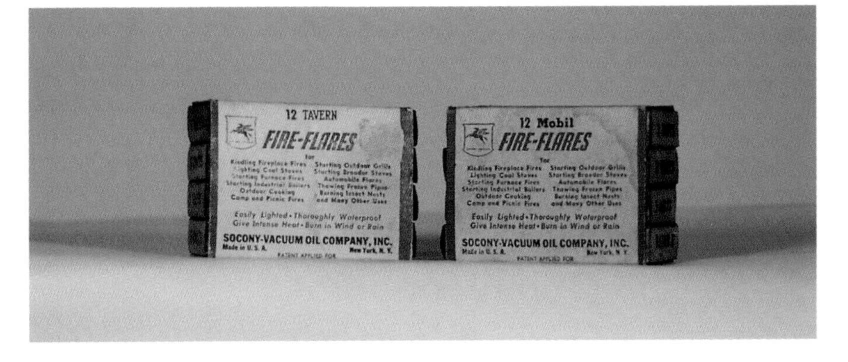

Socony-Vacuum. 12 count Fire Flares. Left, Tavern, $25–35. Right, Mobil, $20–25

Socony-Vacuum. Tavern Novelty Candles, No. 792.
Two Large Choir Boys with box. $25–35

Socony-Vacuum. Tavern Novelty Candles, No. 779.
Four small fawns with box. $25–35

Socony-Vacuum. Tavern Novelty Candles, No. 758.
Large Christmas tree with box. $25–35

Socony-Vacuum. Molded Tavern Birthday
Candles with box. $30–40

Socony-Vacuum. Tavern Novelty Candles, No. 799. Eight Birthday
Boys with box. $25–35

Socony-Vacuum. Tavern Novelty Candles, No. 799 Eight Birthday
Girls with box. $25–35

VI. Giveaways and Miscellaneous

Set of 8 8-oz Friendly Farmer glassware. $35

Set of 8 16-oz 1962 World's Fair glasses. $50

Set of 6 Mobil high ball glasses. $30

Mobil Sports Car glasses. $8–15 each

Team Mobil glasses. $3–6 each.

Mobil NFL glasses with team logos. $3–5

Mobil NFL glasses with team helmets. $2–4

Socony Mobil. Friendly Farmer salt & pepper shaker set. $30–40

Friendly Farmer box with four seasons ash tray or coaster set of 4. $28

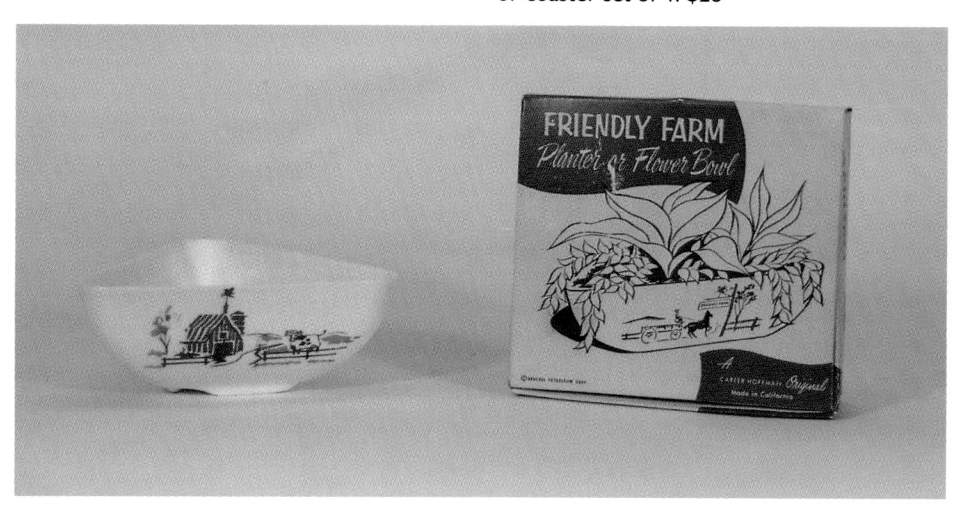

Friendly Farmer box and planter or flower bowl. $15

Mobil soup/salad bowl. $25–30

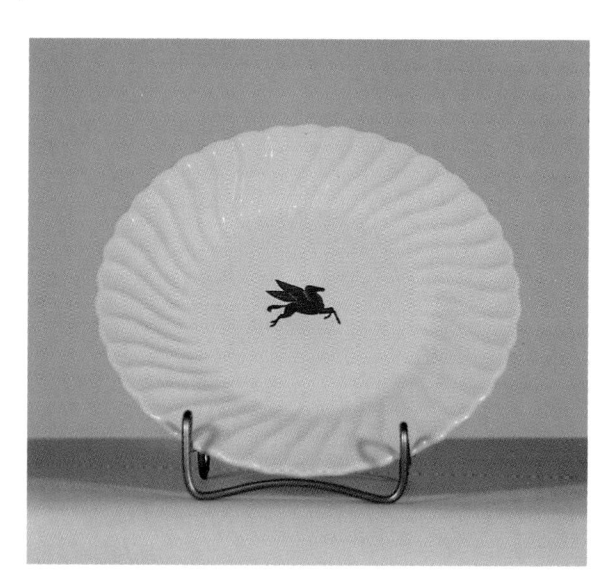

Mobil dessert plate. $25–30

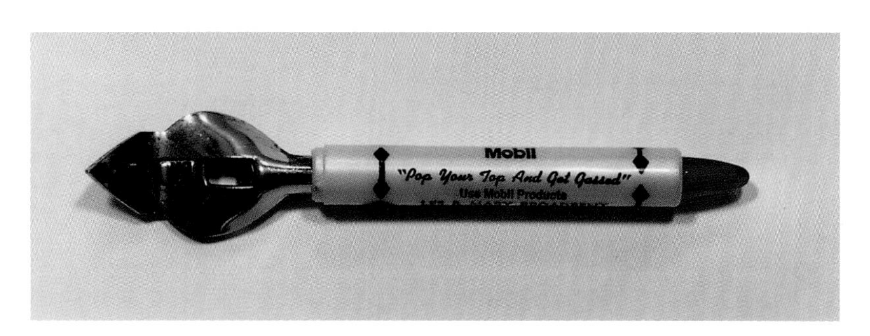

Mobil bottle opener. $10–15

Package of 3 Mobilgas bottle caps. $15–20

Mobil glass baseball bank. $65–80

Bag of marbles given away at Christmas. $75–90

Socony-Vacuum. Mobil Upplerlube bank. $30

Mobiloil cellulose bank. $35

Mobiloil Special Oil can banks. $15–20 each.

Mobil 1 Oil can bank. $20–25

Mobil 1 Oil can transistor radio. $50–65

Socony-Vacuum Oil Co., Gargoyle oil drum paperweight. $50–75

Deck of playing cards with pegasus on back. $8–12

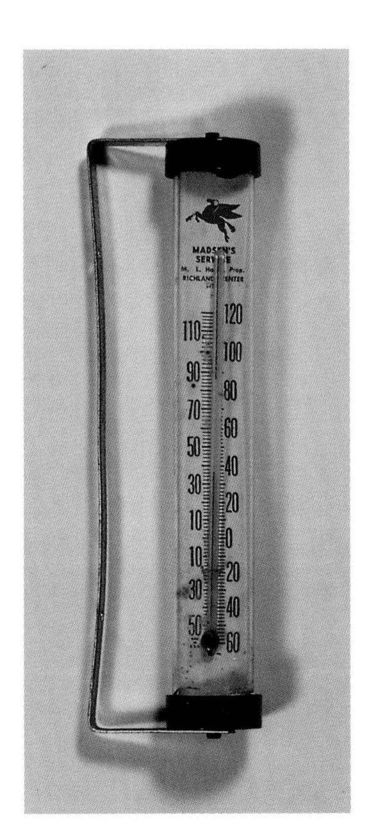

Glass outdoor thermometer. $30–45

Socony Mobil. 2 D-Cell flashlights. $20–25 each

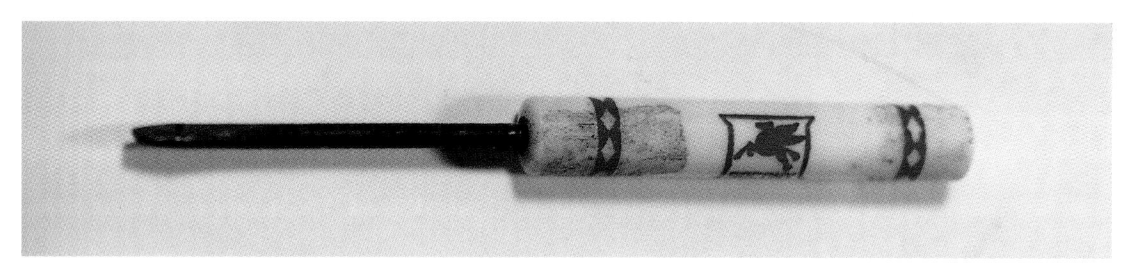

Socony Mobil, screw driver. $6–10

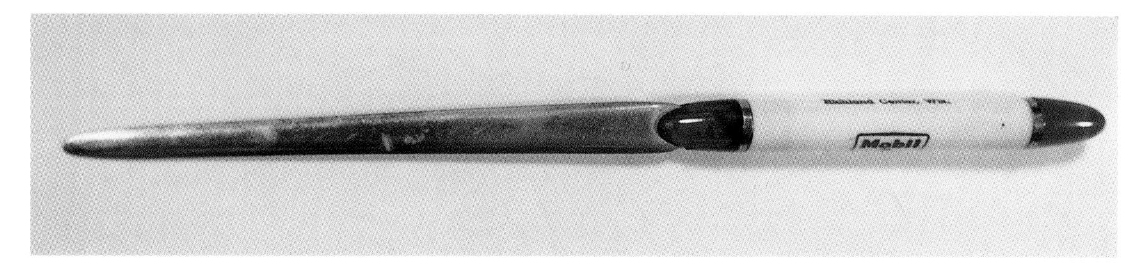

Socony Mobil Letter opener. $8–12

Two patches from an attendant's jacket or shirt. Left, $15–20. Right, $10–15.

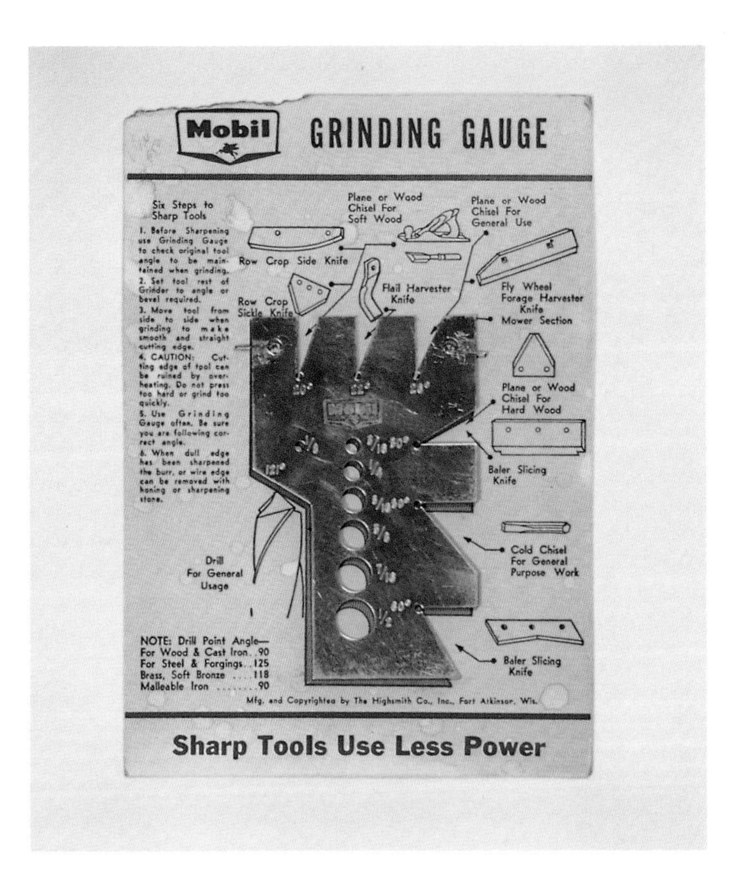

Socony Mobil. Grinding gauge. $15–20

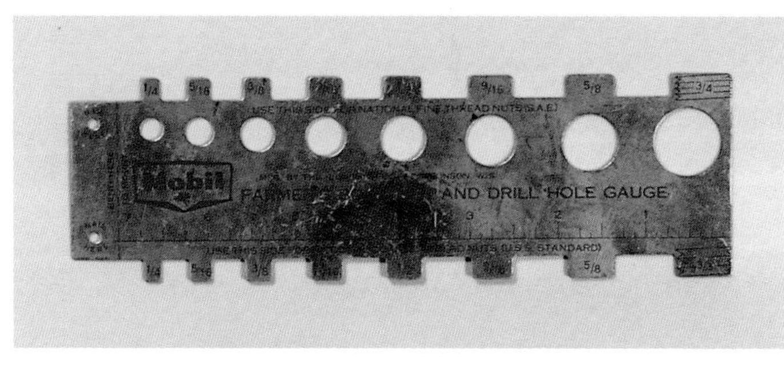

Socony Mobil. Farmer's Nut, Bolt, and Drill gauge. $15–20

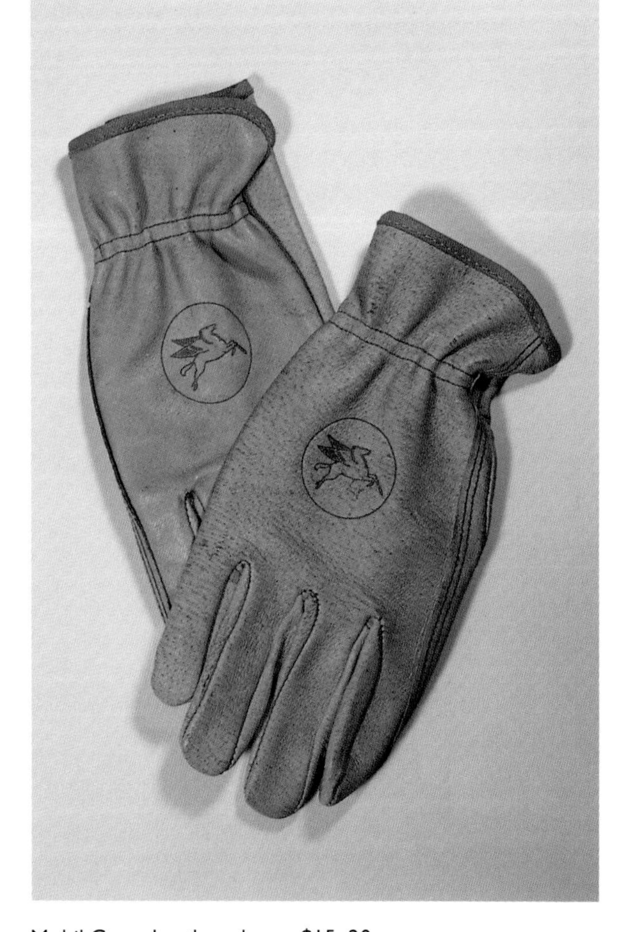

Mobil Corp. Leather gloves. $15–20

Socony-Vacuum. Tractor seat cushion. $65–70

California World's Fair license plate attachment. $150–175

Drive Safely license plate attachment. $90–115

Drive Safely license plate attachment. $85–110

Left to Right: Mobiloil key chain, $4–8. Poker chip key chain with Pegasus, $5–10. Pegasus charm, $4–8. Socony Mobil tape measure key chain, $8–12.

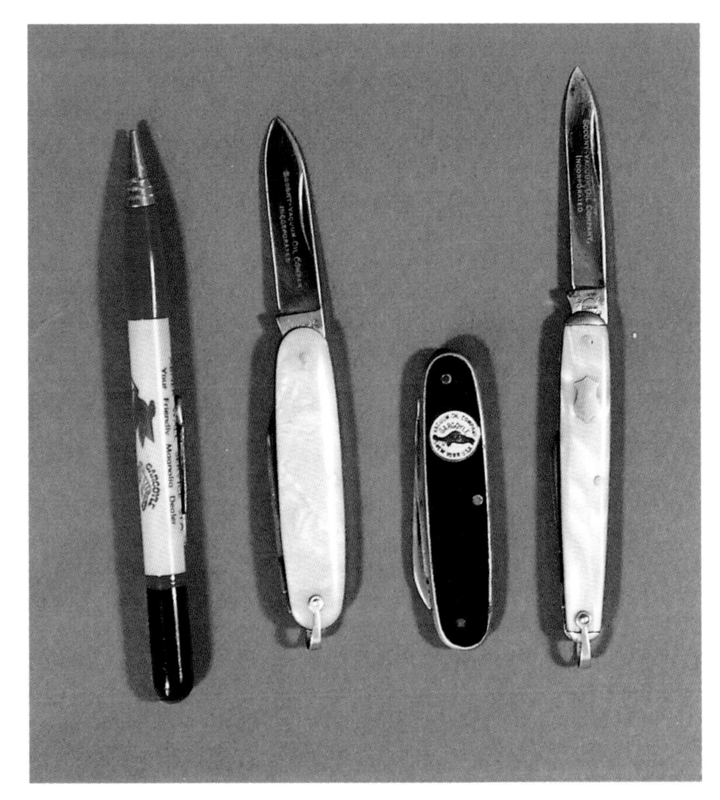

Left to right. Pencil with oil on the top along with the gargoyle, the horse and magnolia on it, $50. Pearl-handled knife made by Remington for Socony-Vacuum, $100. Gargoyle knife, $275. Another pearl-handled knife made by Remington for Socony-Vacuum, $100.

Mechanical Pencils. Left to right: Eight ball, $20–25. Mobilgas/Mobiloil, $15–20. Mobilgas, $15–20. Mobiloil Special with Oil can on top, $45–60. Mobil with emblem on clip, $15–20.

Top left to right: Socony Matchbook cover, Gargoyle lubricant lighter, Red Horse lighter still in wrapper, Mobil Chemical Zippo. Bottom left to right: Vacuum Oil match safe, Magnolia Products, Mobilgas, and Mobil gas pump lighter. $75–175 each.

Socony Mobil Pegasus ash tray. $10–15

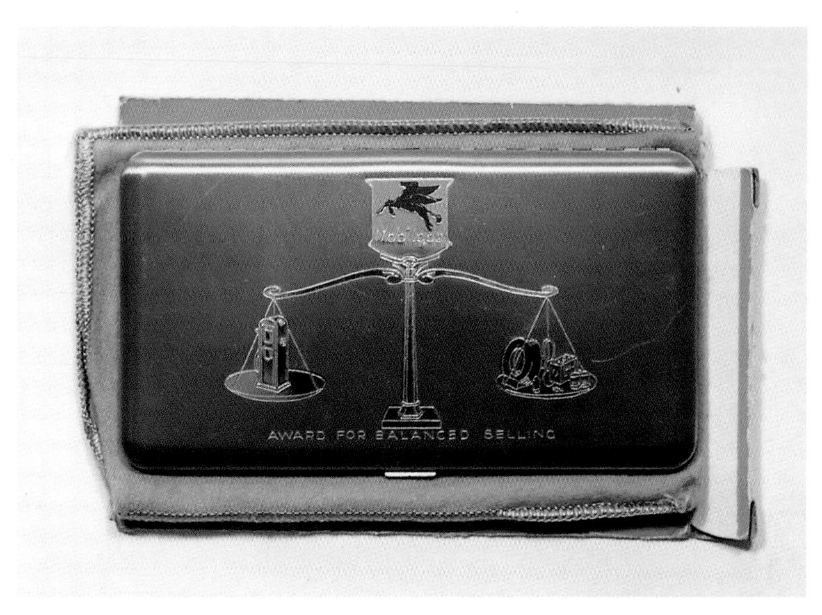

Socony-Vacuum. Brass cigarette case award for "Balanced Selling". $150–175

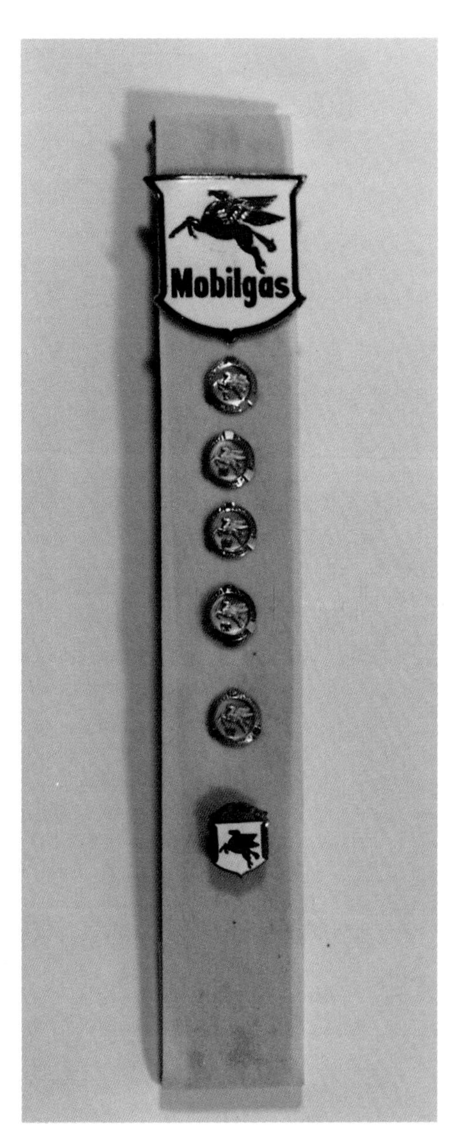

Socony-Vacuum. Top to bottom: Hat pin. Service award pins for 5, 10, 15, 20, and 25 years, with gem stones in the last three. Smaller hat pin. $100–$350 each.

Socony Mobil. Top to bottom: Service award pins for 5, 10, 15, 20, 25 and 30 years with gem stones in the last four. Large hat badge. $100–300 each.

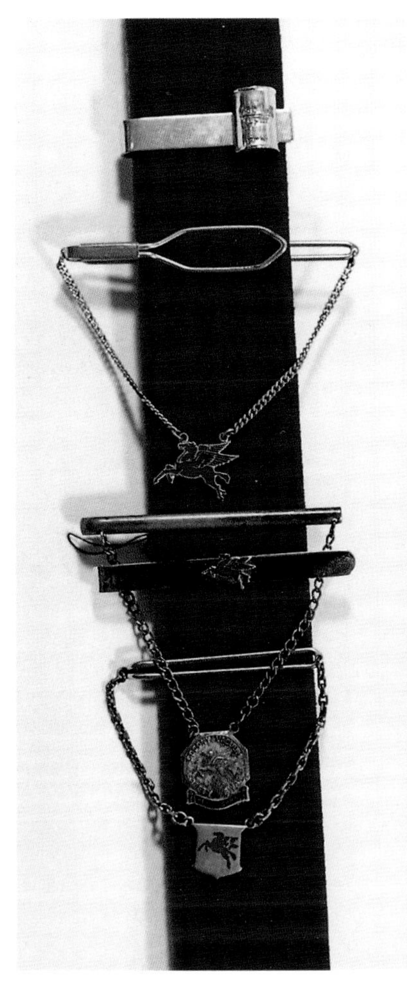

A variety of Socony-Vacuum, Socony Mobil tie bars. $40–75 each.

Another fine collection of Mobil jewelry accessories. $100–350 depending on the piece.

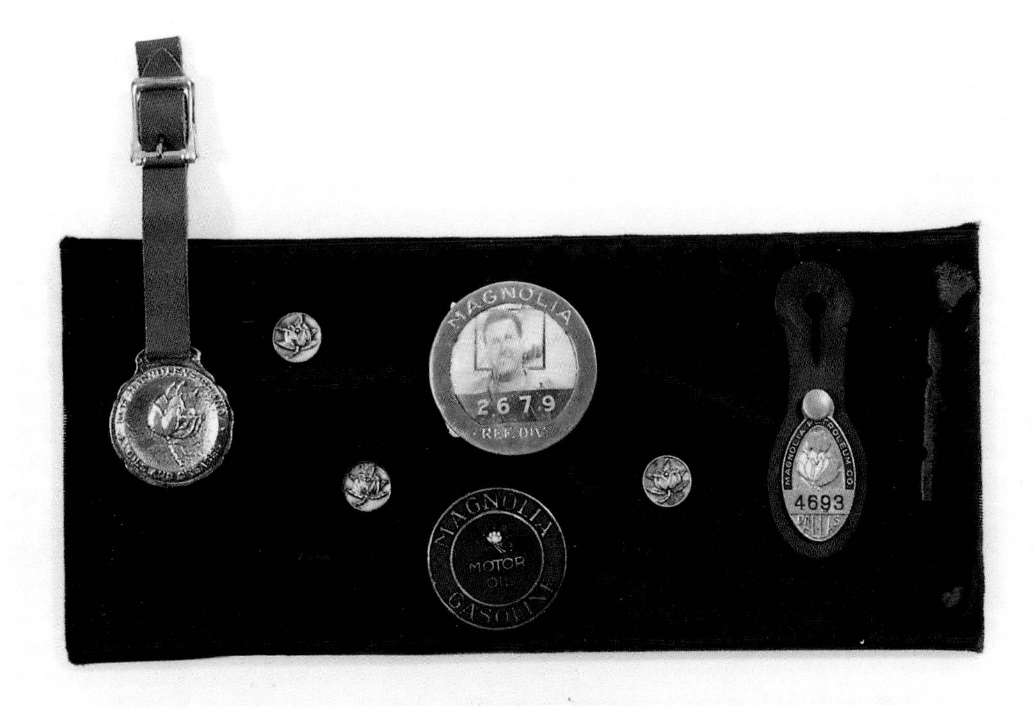

Magnolia smalls collection. Far left, watch fob, Magnoline from Magnolia, $275. Three service award pins, $125 ea. Center top, refinery employee badge, $150. Center bottom, hat badge. Far right, employee button badge. The hat badge and the button badge are priceless to the owner.

Vacuum Oil Co. Gargoyle salesman's kit with four of the six different jars. $750–1000

Vacuum Oil Co., sample kit jar of valve oil. $75

Vacuum Oil Co., sample kit jar of Mobiloil "C". $75

Socony Mobil. 3" tube of Mobil grease Special with (Moly) printed on back. $12–15

Front of the 3" tube.

Gargoyle hood ornament. $250–300

Vacuum Oil Co., Gargoyle Marine Oil clock with a bakelite case and a patent date of March 29, 1927. 2 1/2" face. $500 and up.

VII Toys

Smith-Miller truck 22" long. this truck has "Mobilgas" on one side with a door that has a retractable hose, and the other side has "Mobiloil" on it. $495–545

Smith-Miller, R-Model mack with pup. $850

Tin-stamped truck with static chain made in Japan. $125–150

10" 2-piece Mobilgas truck with friction motion, tin-stamped, made in Japan. $105–135

Tin-stamped truck made in Japan. $95–135

10" Mobilgas truck with friction motion, tin-stamped, made in Japan. $125–150

11" Mobilgas truck with friction motion, tin-stamped, made in Japan. $125–150

Tin-stamped Mobilgas truck with friction motion,
made in Japan. $95–135

3" Mobilgas truck, tin-stamped with friction motion, Line MAR toys, made in Japan. $50–65

11" Mobilgas truck, tin-stamped with friction motion, made in Japan. $175–200

12" Mobilgas truck, tin-stamped with windout hose in back. $185–225

2-piece 9" Mobil truck, by Tootsie Toy of Chicago U.S.A. $65–80

2-piece 4 1/4" truck, by Tootsie Toy of Chicago U.S.A. $45–70

4 1/4" Mobil oil truck, tin-stamped with friction motion, made in Japan. $45–70

2-piece 9" Mobilgas truck, by Tootsie Toy of Chicago U.S.A. $65–80

9" Mobilgas truck with box , tin-stamped with friction motion, made in Japan. $325–375

4 1/2" Mobilgas truck, made in England by Dinky Toy. $45–70

6 1/2" Mobilgas truck, plastic body with a tin-stamped tank, friction motion, made in Japan. $25–40

2-piece 10" Mobilgas plastic truck with friction motion, made in Hong Kong. $25–40

5" Mobil plastic truck made in Hong Kong. $20–35

3 3/4" Mobil truck with plastic tank and metal cab cover, made in England. $20–35

2-piece 10 1/4" Mobil tanker with plastic tank and metal cab made by Ertl in Hong Kong. $30–50

2-piece 22 1/2" Mobil Oil Corp. Fairfax, VA, Semi tractor and trailer, made by Ertl in Dyersville, Iowa. $50–75

2-piece 23" Semi with plastic tank, tank has a working pump and metal body, made by Ertl in Dyersville, Iowa. $85–135

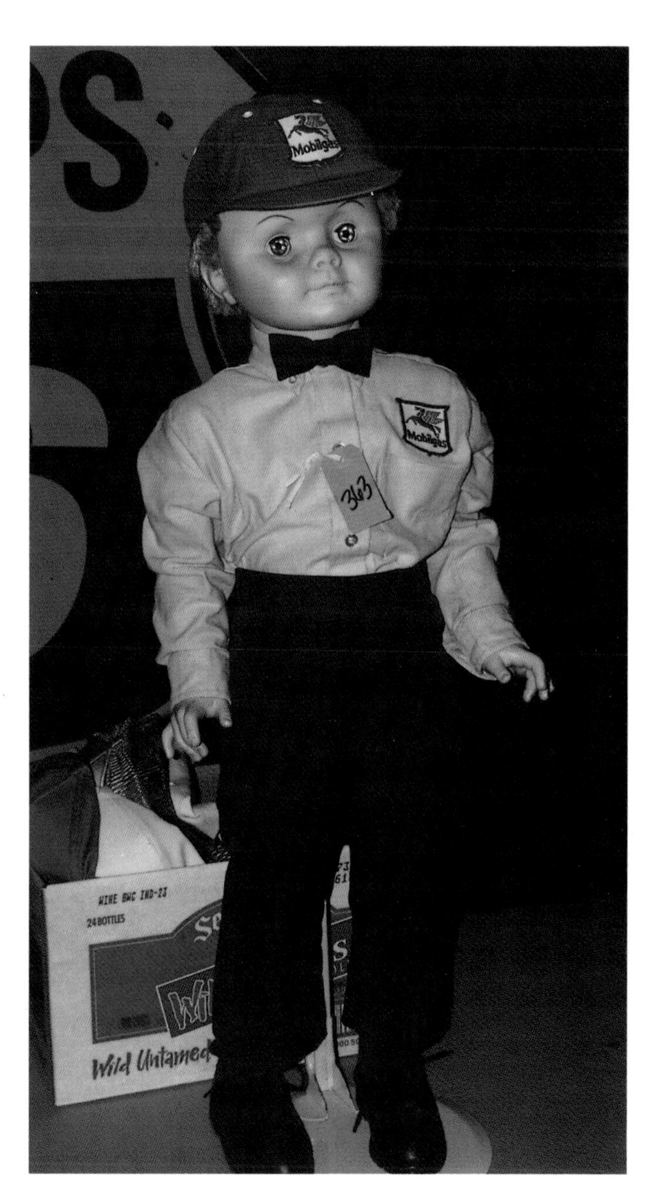

3' tall attendant doll with uniform. $35–50

VIII. Paper

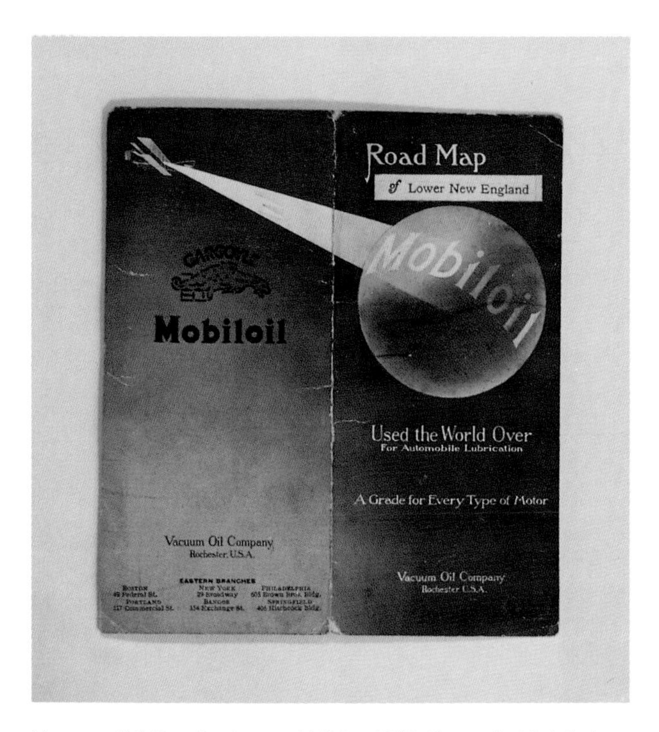

Vacuum Oil Co., Rochester U.S.A., 1912 Gargoyle Mobiloil map. This map is believed to be the oldest oil company road map; note the Wright brothers plane on the back of the map. $600

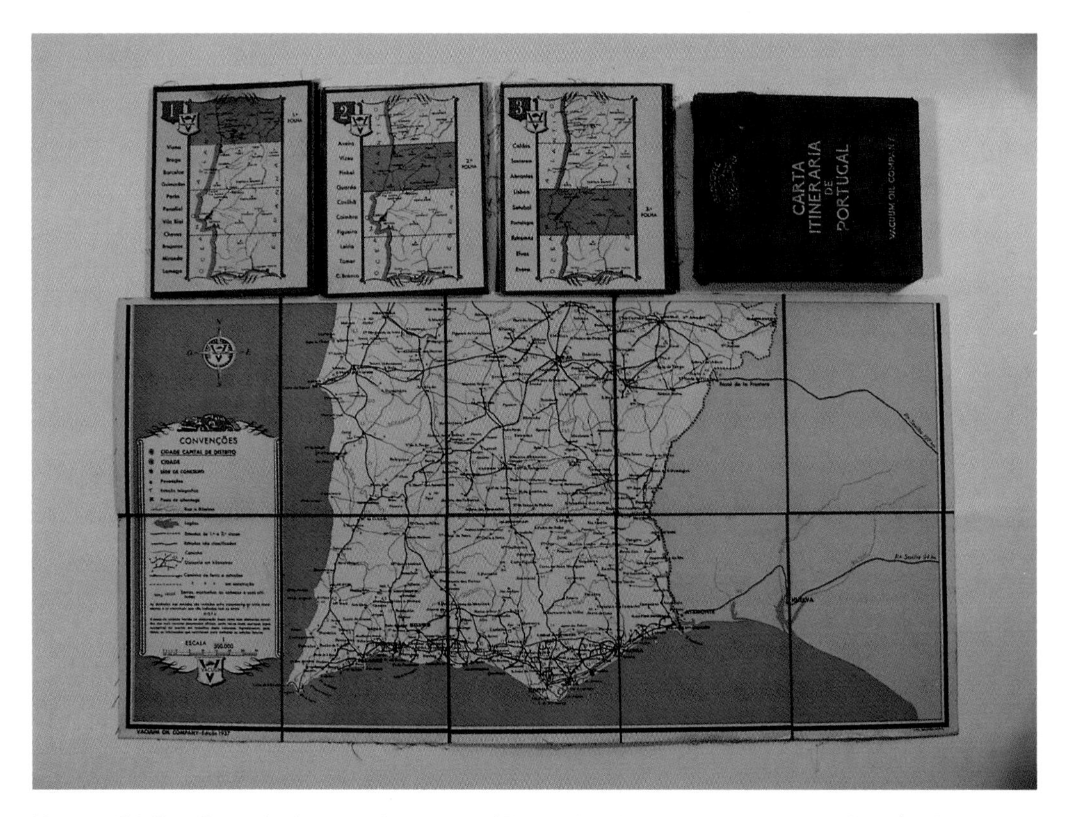

Vacuum Oil Co., Gargoyle 4-section linen map of Portugal with carrying case, dated 1937. $500

Standard Oil Co. of N.Y., Socony, 1926 road map of New York. $25–30

Socony-Vacuum street guide for Chicago IL, copyright 1949. $15–25

Wadhams 1936 road map of Wisconsin. $20–30

Wadhams 1939 road map of Wisconsin. $15–25

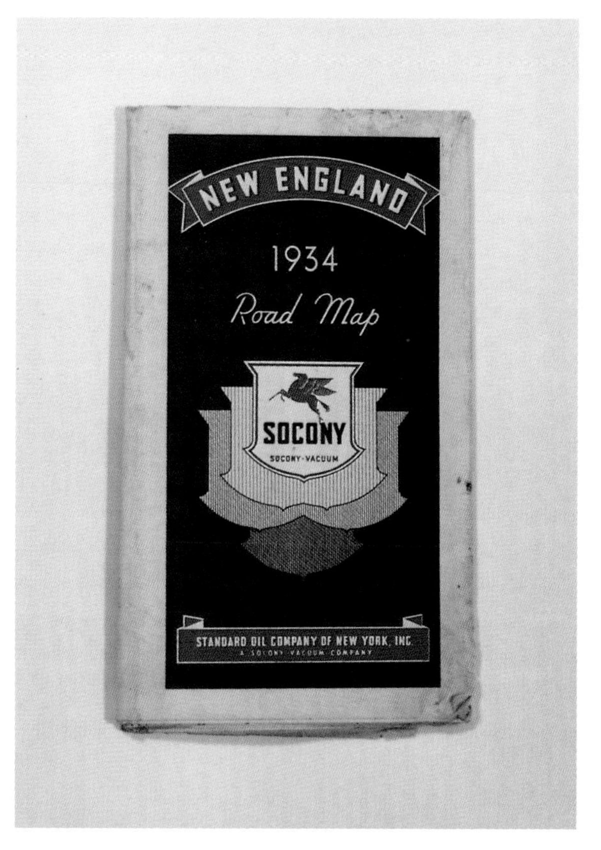

Standard Oil Co. of New York, Socony-Vacuum, 1934 road map of New England. $30–40

Standard Oil of New York, Socony-Vacuum 1937 road map of the Finger Lakes Region of New York. This map is opened to read "Travel the Route of Friendly Service". $25–35

Socony-Vacuum Oil Co., Inc. 1940 road map of Westchester & Putnam Counties with adjoining territory. The interior of this map has The New York World's fair printed on it. $15–25

Socony-Vacuum Oil Co., Inc.1935 road map of Minnesota and adjoining states. $15–25

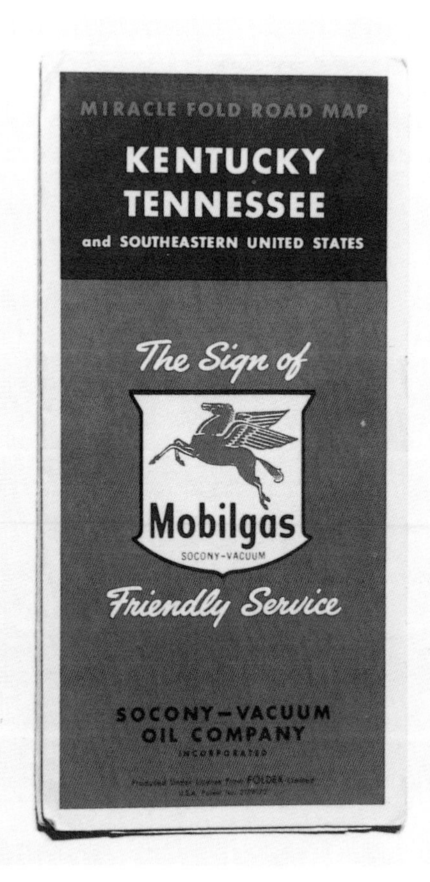

Socony-Vacuum Oil Co., Inc. 1950 census road map of Kentucky & Tennessee. $10–20

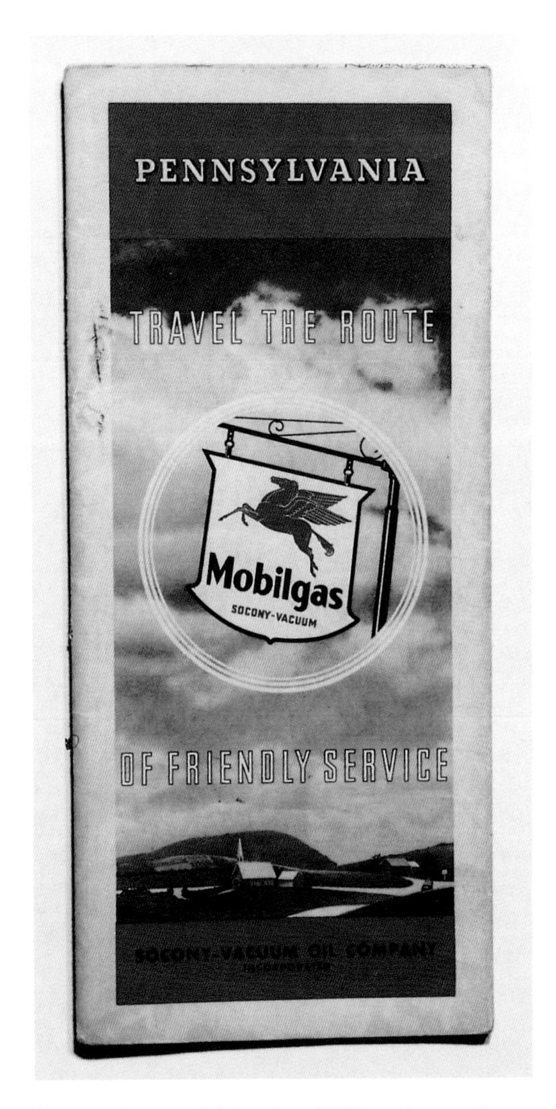

Socony-Vacuum Oil Co., Inc. 1937 road map of Pennsylvania. $20–30

Socony-Vacuum Oil Co., Inc. early 1940s road map of Wisconsin, Upper Michigan, and adjoining states. $15–25

Magnolia Petroleum Co., a Socony Mobil Co., mid-1950s road map of the Southwest. $10–20

Socony-Vacuum Oil Co., Inc. Lubrite division. 1934 road map of Missouri. $20–30

Socony-Vacuum Oil Co., Inc. White Eagle Division. 1933 road map of North & South Dakota. $20–30

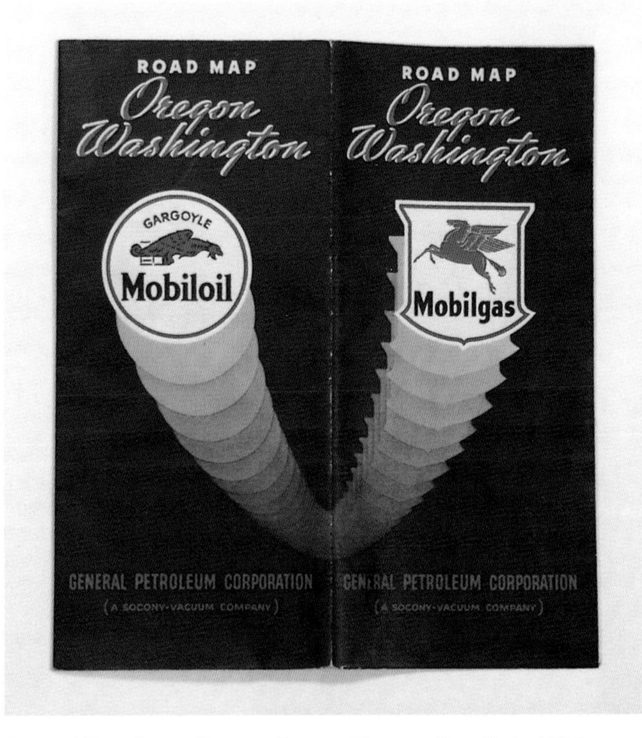

Socony–Vacuum Oil Co., Inc., White Eagle Division. 1930 road map of North & South Dakota. $20–30

General Petroleum Corp., a Socony-Vacuum Co., Early 1930s road map of Oregon & Washington. $20–30

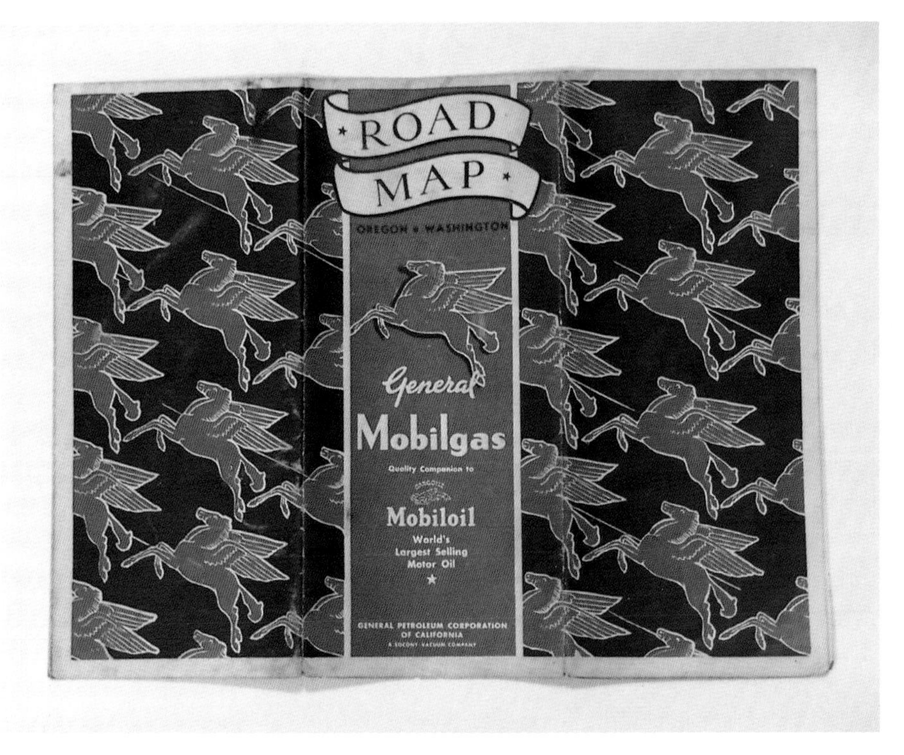

General Petroleum Corp of California a Socony Vacuum Co., early 1930s road map of Oregon & Washington. $20–30

General Petroleum Corp., early 1930s "Mobilize with Mobilgas" Mobiloil road map of Oregon & Washington. $25–35

General Petroleum Corp., a Socony-Vacuum Co., Early 1950s road map of San Francisco and East Bay Cities. $10–20

General Petroleum Corp., a Socony-Vacuum Co., 1940s road map of California, Nevada, Arizona and Utah. $10–20

General Petroleum Corp., a Socony-Vacuum Co., 1930s Los Angeles City Map. $20–30

General Petroleum Corp, a Socony Mobil Co., 1957 Pacific Coast Cruising Guide. $15–25

General Petroleum Corp., a Socony Mobil Co., 1956 census California road map. $10–20

General Petroleum Corp., a Socony Mobil Co., 1950s Seattle and Portland road map. $10–20

Socony Mobil Oil Co., Inc. 1950s New York road map. $10–20

Socony Mobil Oil Co., Inc. 1950s Southern New England road map. $10–20

Socony Mobil Oil Co., Inc. 1950s Minnesota road map. $10–20

Mobil Oil Corp. 1960s Central & Western United States road map. $5–10

Mobil Oil Corp. 1970s Iowa road map. $3–6

Socony-Vacuum Oil Co., White Eagle Division map rack. Without the maps, $400.

Mobilgas map rack. Without the maps, $100–125.

Mobil Map rack, $325–400 without maps.

Socony-Vacuum Oil Co., Wadhams Division. 1942 Flying Red Horse almanac. $55–70

Socony-Vacuum Oil Co., Wadhams Division. 1943 Flying Red Horse almanac. $55–70

Socony-Vacuum Oil Co., Wadhams Division. 1944 Flying Red Horse almanac. $55–70

Socony-Vacuum Oil Co. 1945 Flying Red Horse almanac. $55–70

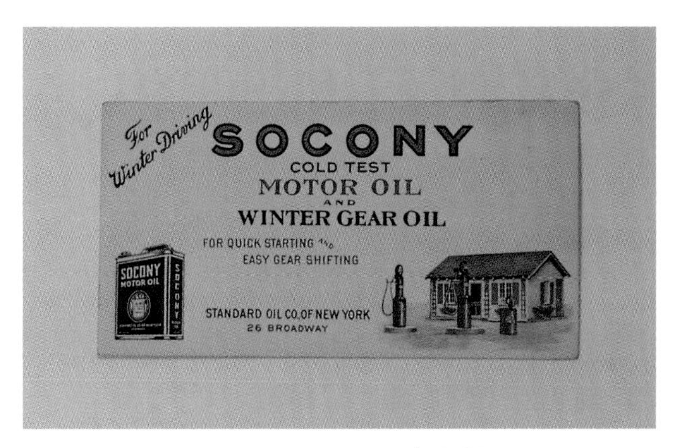

Standard Oil Co. of New York blotter. $30–35

Socony-Vacuum Oil Co. Reminder of last service card. $15–20

Variety of Socony Mobil Oil Co. Inc. blotters. $ 12–18 each.

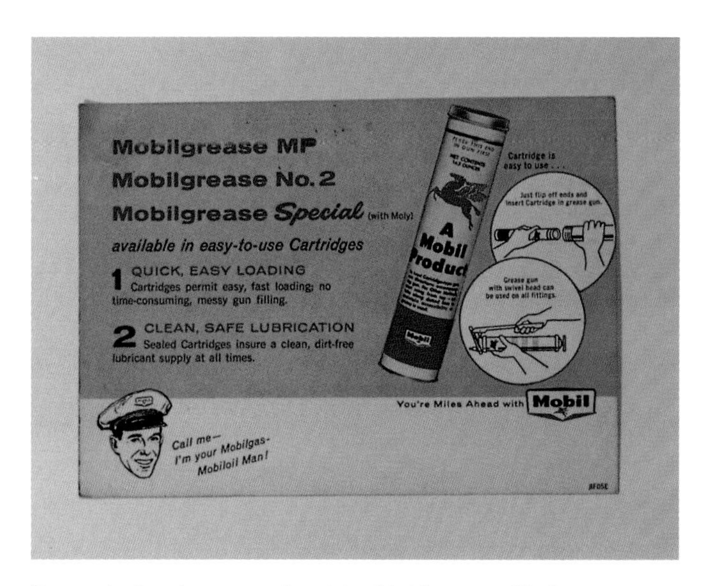

Postcard advertisement, advertising Mobil grease. $5–8

Socony Mobil Oil Co. Inc. Three-fold advertisement for three different products available for sale. $8–12

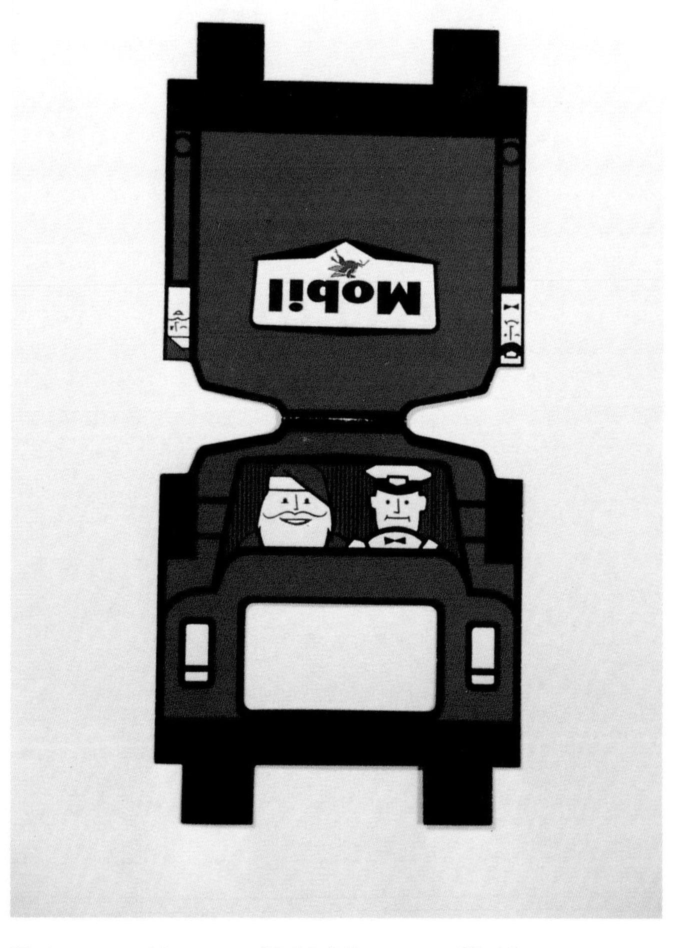

Christmas card from your Mobil delivery man. $5–10

Advertisement fan, front & back. $50–60

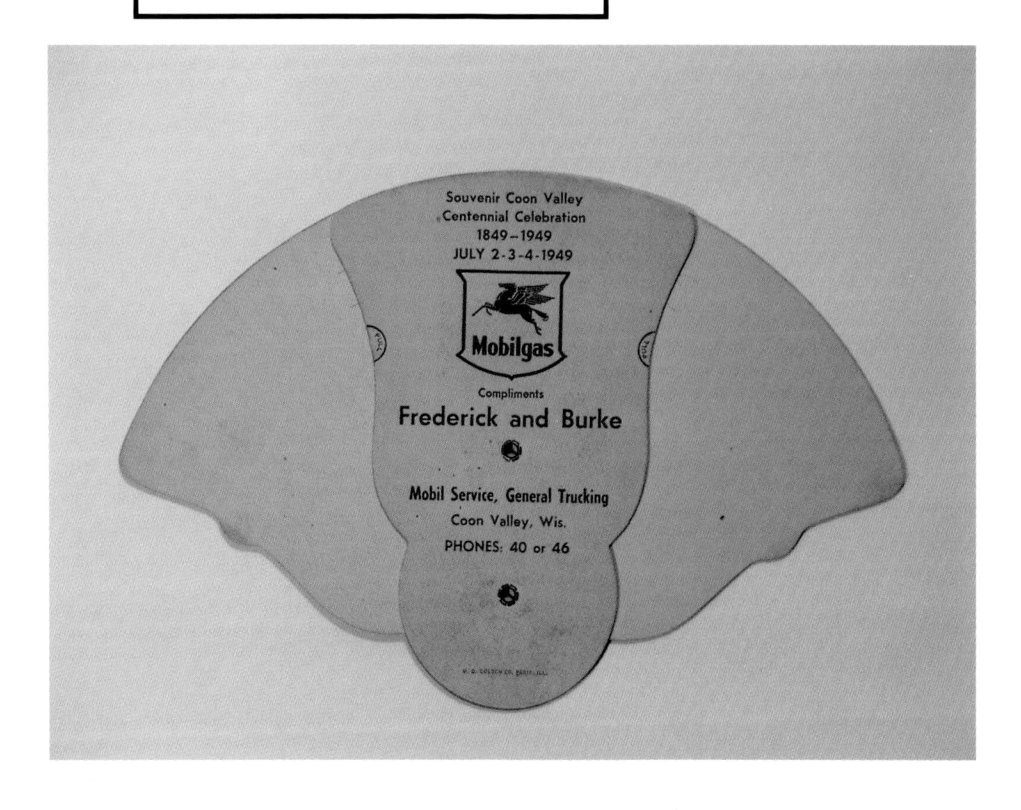

1953 Mobilgas pocket calendar. $15–20

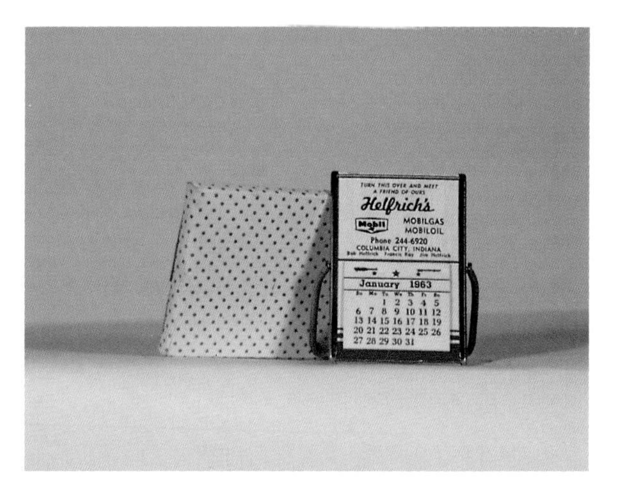

Women's 1963 vanity calendar with mirror on the other side, with box. $25–30

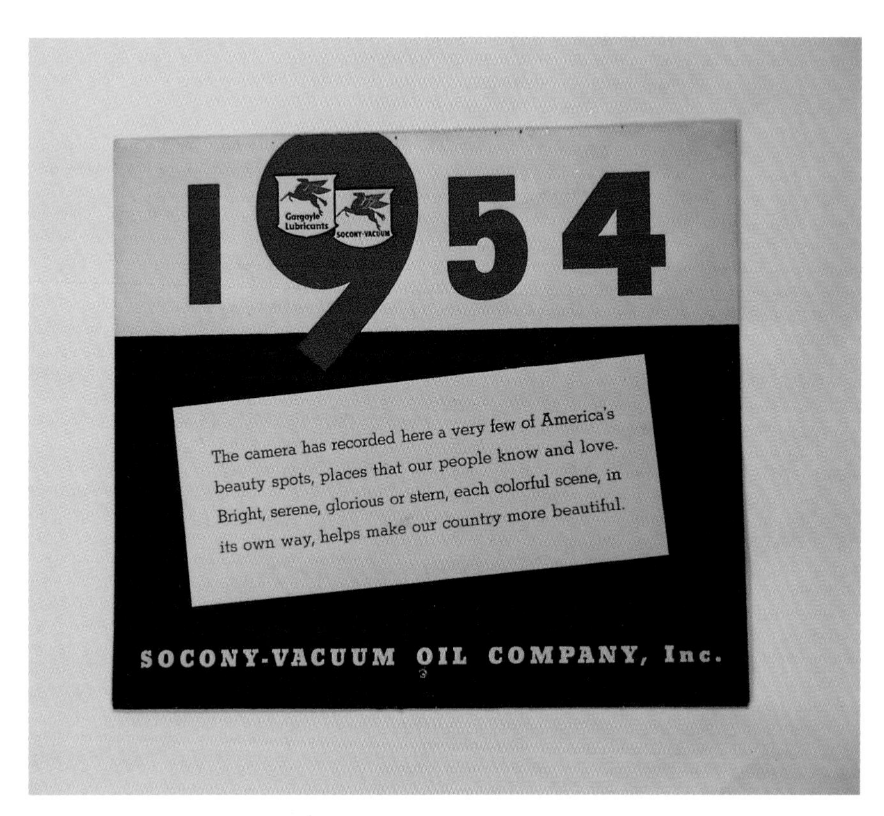

Socony-Vacuum Oil Co., Inc. 1954 wall calendar. $15–20

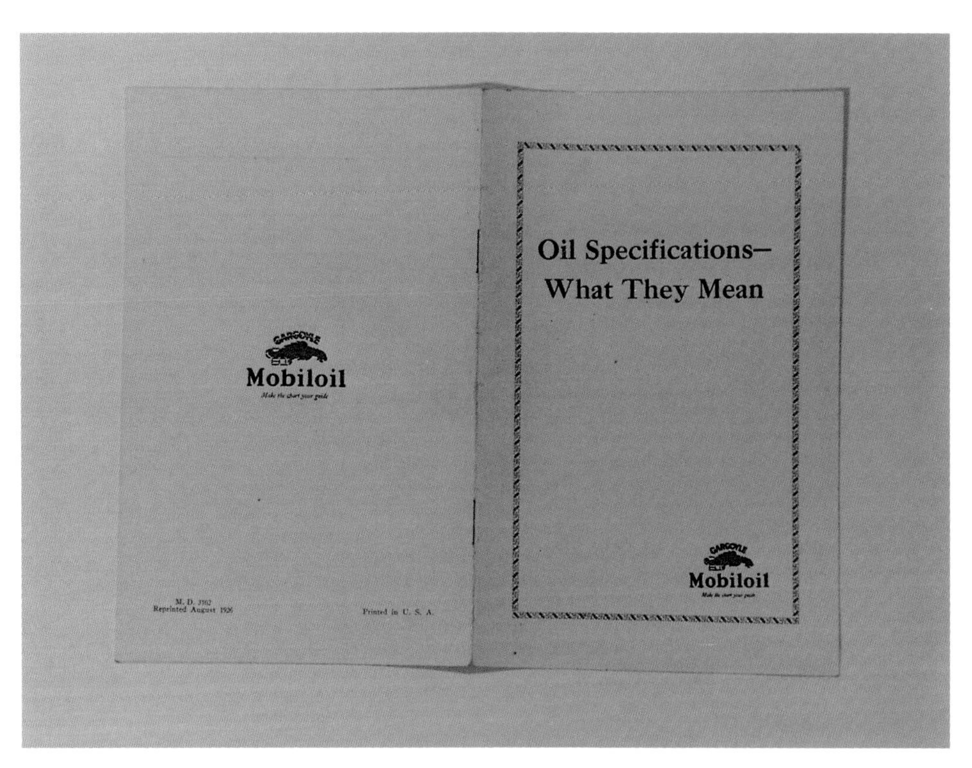

Gargoyle Oil Specification booklet reprinted in 1926. $15–20

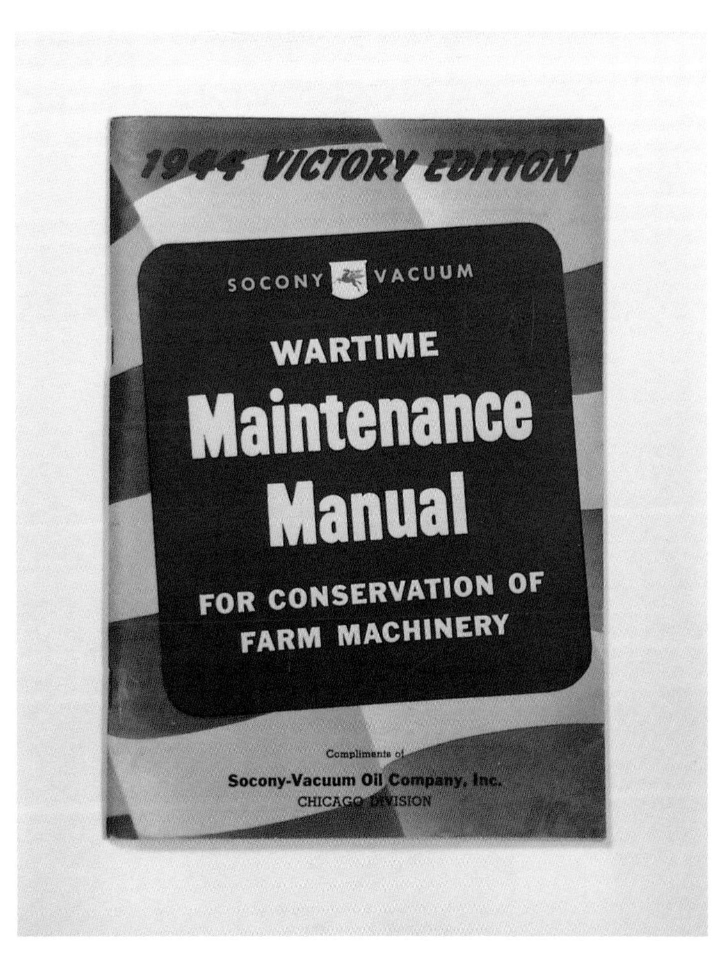

Socony-Vacuum Oil Co., Inc. 1944 Victory Edition Maintenance
Manual for your farm machinery. $55–70

Socony-Vacuum Oil Co. service record for your vehicle. $10–15

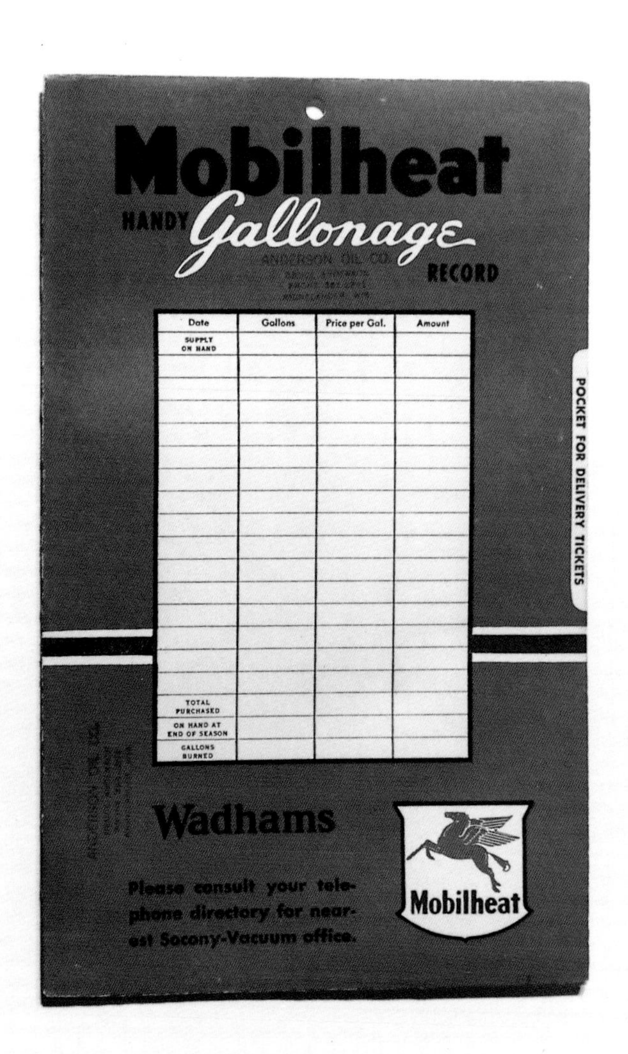

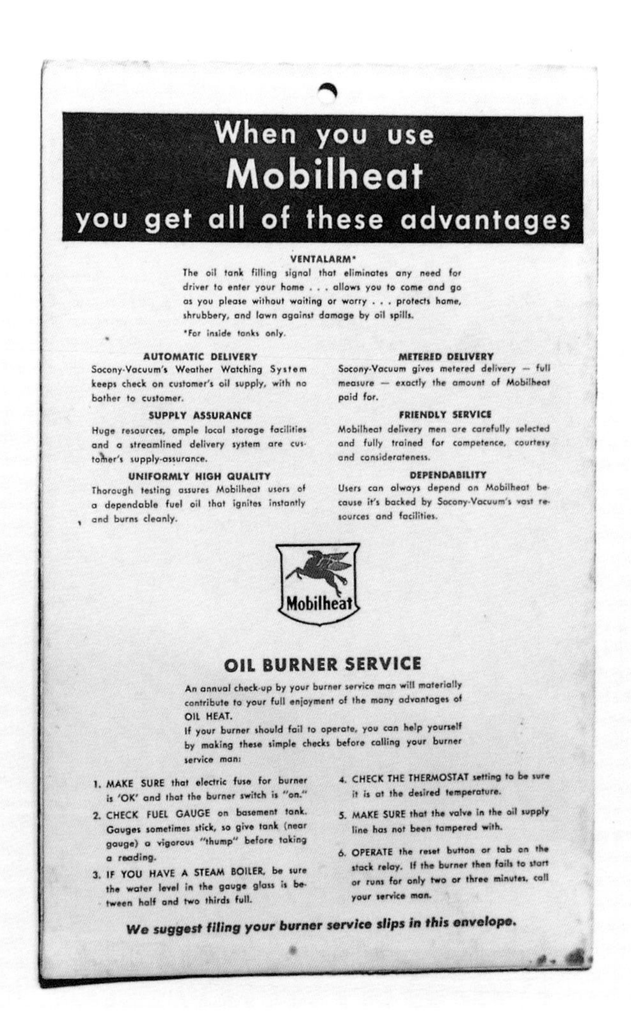

Wadhams Mobilheat gallonage record for home heating fuel oil, front & back. $25-30

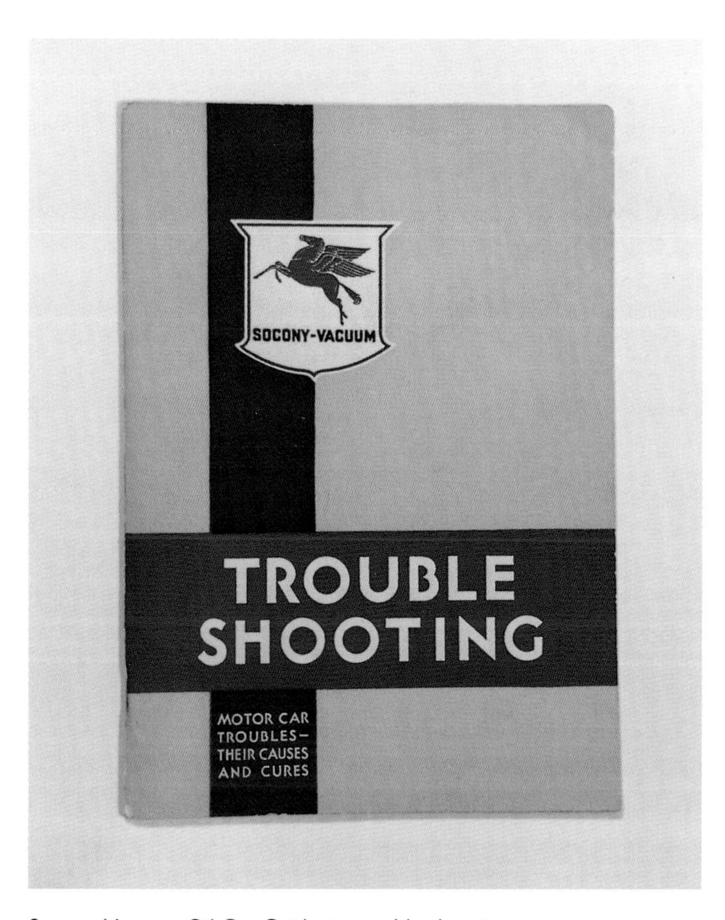

Socony-Vacuum Oil Co. Guide to trouble shooting vehicle problems. $55–70

Socony-Vacuum Oil Co. "Better Homemaking" by Effa Brown. $30–40

Socony-Vacuum Oil Co., Wadhams division. 1003 household hints and worksavers. $55–70

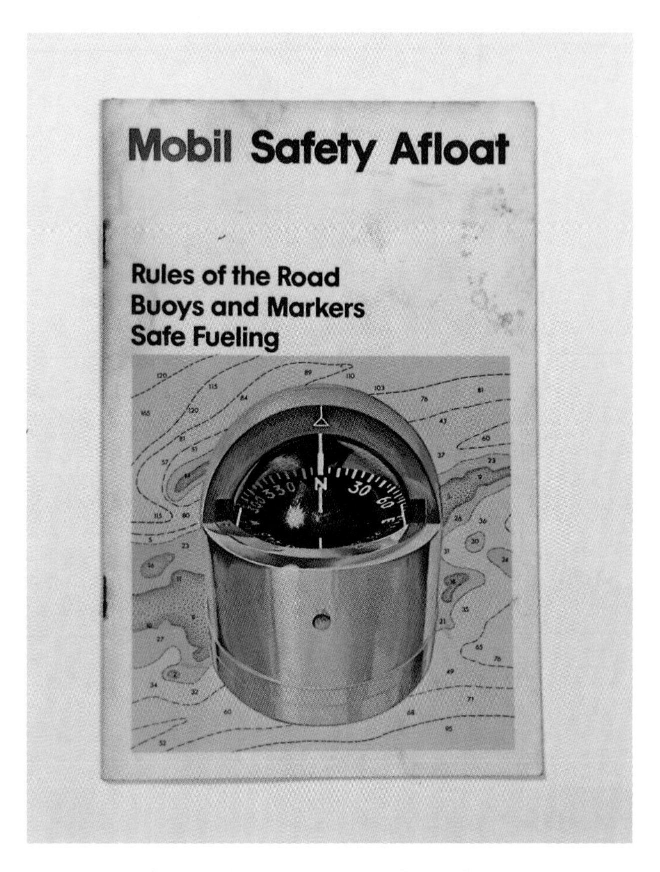

Mobil Corp. Safety Afloat: Rules of the Road, Buoys and Markers, Safe Fueling. $15–20

Socony-Vacuum Oil Co., Inc. Advertisement proof to appear in the October 26, 1940 *Saturday Evening Post* and the October 28, 1940 *Life* magazine. $ 15–20

Socony-Vacuum Oil Co., Inc. 1940s advertisement from *Life* magazine. $6–9

Socony Mobil Oil Co., Inc. 1955 advertisement, "Whatever they Dream Up Next." $5–7

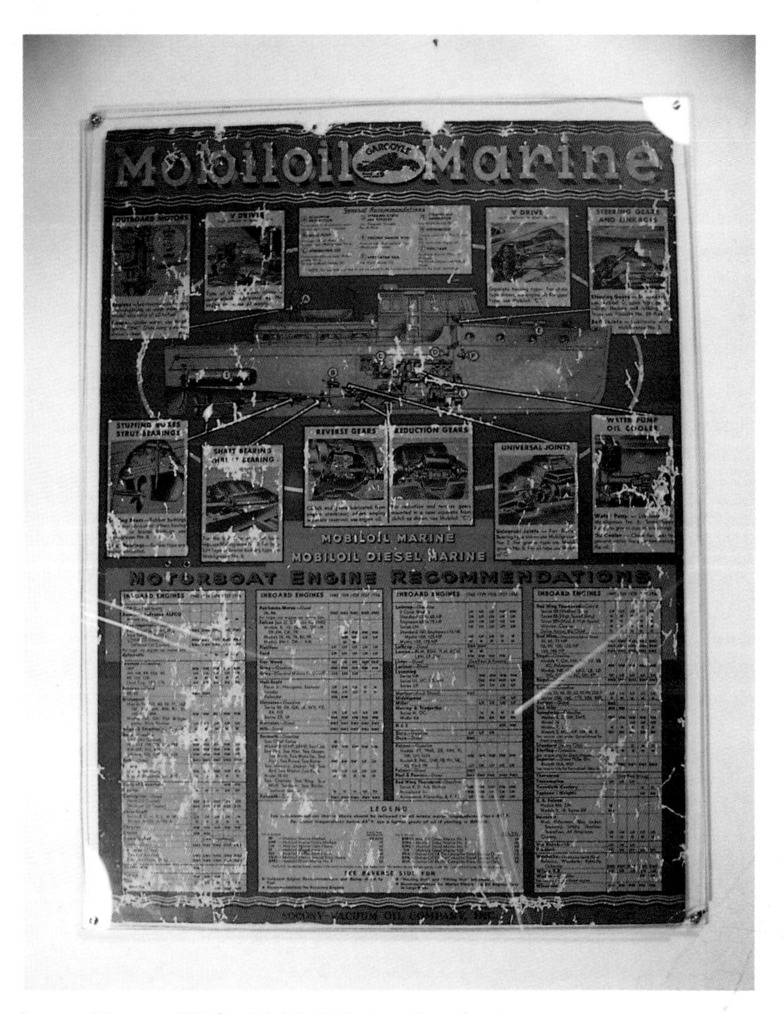

Socony-Vacuum Oil Co. Mobiloil Marine chart for the motorboat engine. $55–70

Socony-Vacuum Oil Co., Inc. 1940s advertisement from *Life* magazine. $6–9

Socony-Vacuum Oil Co., Inc. Advertisement proof to appear in the June 17, 1940 *Life* magazine. $15–20

Socony-Vacuum Oil Co., Inc. 1940s advertisement from *Life* magazine. $6–9

Vacuum Oil Co. *The Gargoyle* newsletter, December 1920. $40–50

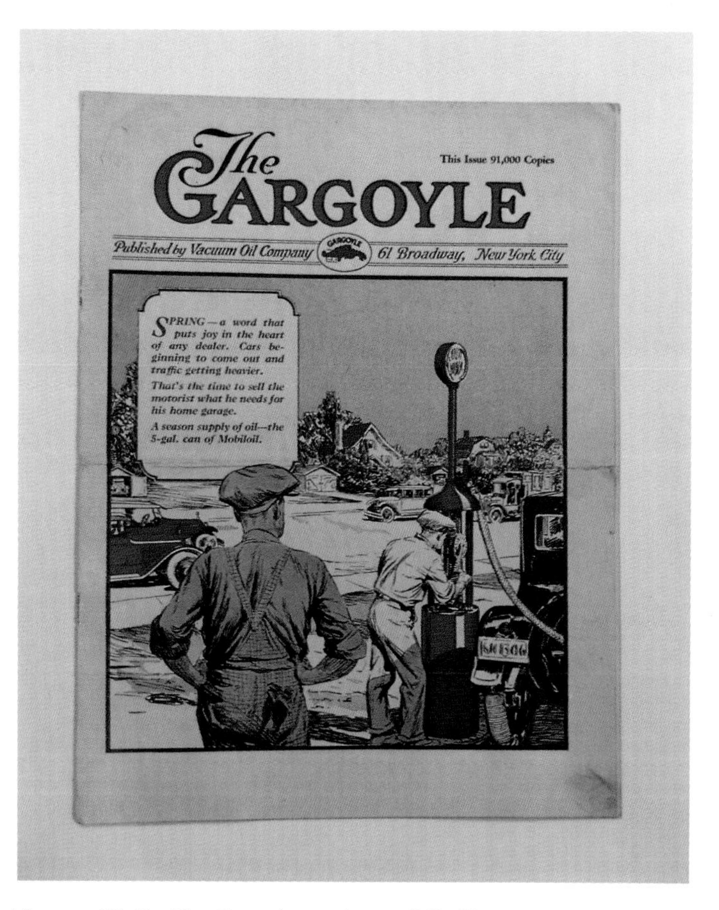

Vacuum Oil Co. *The Gargoyle* newsletter. $40–50

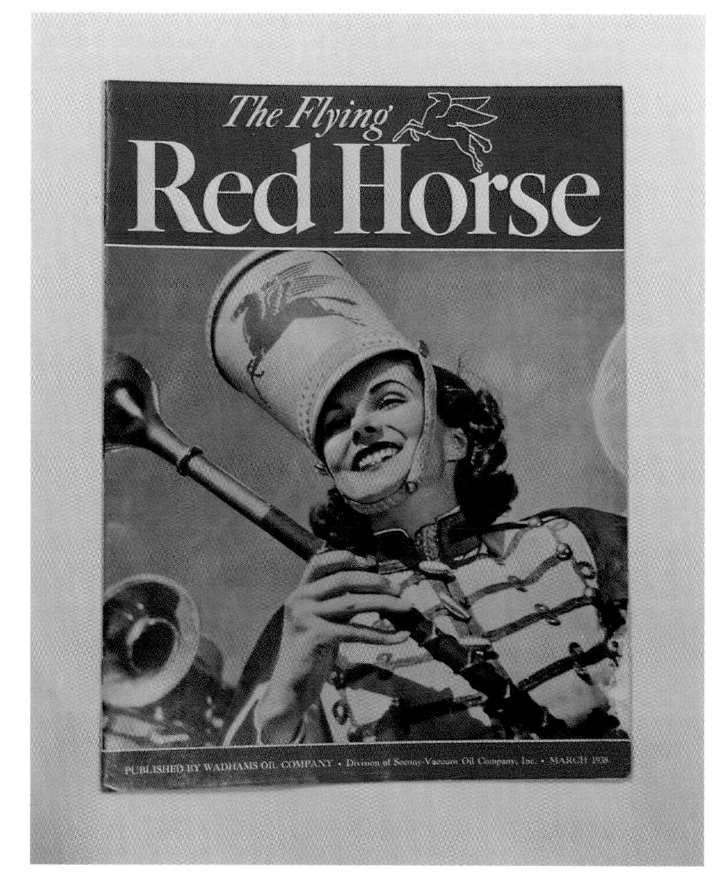

Wadhams Oil Co., a division of Socony-Vacuum. *The Flying Red Horse* magazine, March 1938. $40–50

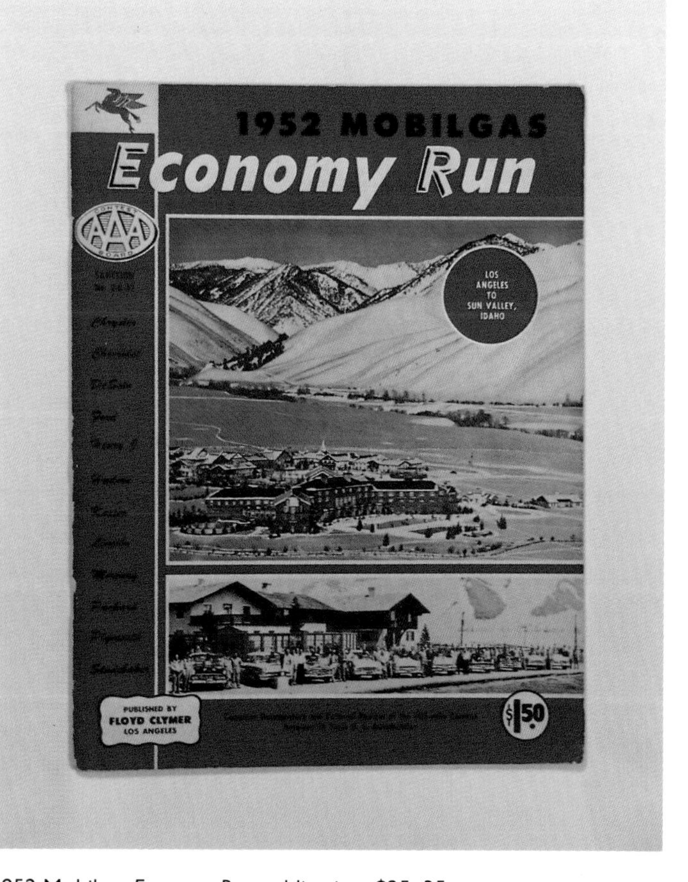

1952 Mobilgas *Economy Run* publication. $25–35

SOCONY-VACUUM OIL COMPANY
INCORPORATED
26 BROADWAY
NEW YORK

February 5, 1935.

TO THE STOCKHOLDERS:

The Board of Directors today declared a dividend of 15 cents per share, payable March 15, 1935, to stockholders of record February 20, 1935. The Directors also decided to consider dividend action in the future semi-annually in February and August, for March and September payment. This policy should enable the Directors to act with better knowledge of the Company's financial requirements and the trend of earnings each year. The next regular meeting for dividend action will be in August, this year.

Earnings for 1934 can not be accurately stated until accounting is completed but are estimated to be more than in 1933. Current asset position was somewhat reduced, but continues strong. Cash and marketable securities at the end of the year were approximately $54,000,000, it being impossible to state the exact amount until reports from all world operations are received. Necessary capital expenditures, always large, were particularly great in 1934, and it is expected that they will be unusually heavy for some time. This is due in part to being unable to defer any longer important improvements and replacements and in part to trends in the industry requiring new investment in plants and production to protect the future of your business.

Your Directors believe that prudent and conservative management requires the change in dividend policy of which you are now informed. Future dividends will depend upon the earnings and current asset position with relation to the need for maintaining the strength of your business, and today's action is not to be considered as indicating any regular dividend basis.

During the past year, consumption of petroleum products in the world increased and your Company increased its sales. Potential oversupply of crude oil and difficulties of marketing competition have operated to prevent satisfactory prosperity for the industry as a whole. The Directors will endeavor to keep your Company in a position to benefit from any future improvement in the industry's condition.

BY ORDER OF THE BOARD OF DIRECTORS

Chairman of the Executive Committee

Letter to the stockholders from Socony-Vacuum Oil Co., 1935. $8–12

TANK WAGON CONSUMERS GASOLINE CONTRACT

Contract, made this.........................day of.........................., 193..... between SOCONY-VACUUM OIL COMPANY, INCORPORATED, a New York corporation, having a place of business at ...., hereinafter called Seller, and .... of .... jointly and severally, if more than one, hereinafter called Buyer.

FIRST: *Sale.* Seller hereby agrees to sell and deliver and Buyer agrees to buy, receive and pay for during the term hereof and in accordance with the provisions of this Contract the quantities set forth below of the products named below, and at the prices hereinafter specified. All products purchased hereunder are for Buyer's own consumption only and not for resale.

SECOND: *Products.* In the performance of their respective obligations, Seller will deliver and Buyer will receive and pay for the following grades and/or brands of products of the kind and quality marketed by Seller at time and place of delivery, all as determined by Seller:

If the products covered hereby are not described by brand names or other specific designations, then the products to be delivered and accepted hereunder shall be such of said classes of products as Seller shall from time to time be selling to other customers similar to Buyer in Buyer's territory, all as determined by Seller. Any octane ratings referred to in this contract shall be determined by the A. S. T. M. method prevailing at the time of shipment.

THIRD: *Quantities.*

FOURTH: *Prices.* The price for each grade or brand shall be Seller's tank car price for commercial consumers in effect at time of delivery and place from which delivery is made as posted by Seller at such place, plus two and a half cents (2½c) per gallon except that on each delivery of less than 100 gallons the price shall be Seller's service station price in effect at time and place of delivery as posted at the place from which delivery is made.

Discounts from said prices shall be as follows:

DELIVERIES PER MONTH	DISCOUNTS PER GALLON
100 to 3,000 gallons	None
3,000 to 10,000 gallons	One half cent
10,000 to 20,000 gallons	One cent
20,000 to 40,000 gallons	One and a half cents
40,000 and over	Two cents

In computing deliveries for discount purposes, deliveries of all grades or brands shall be combined provided, however, that deliveries of less than 100 gallons each shall not be included for discount purposes. No discount shall be paid or allowed at time of delivery but shall be due and payable within twenty-five days from the end of the month in which deliveries are made. The term "month" as used in connection with prices and discounts means the calendar month. Discounts for deliveries during fractional parts of calendar months at the commencement and ending of the contract period shall be computed proportionately. If at the end of the yearly period the aggregate of discounts computed in accordance with the foregoing schedule and on the basis of average monthly deliveries exceeds the aggregate of discounts computed on the actual monthly deliveries, Buyer shall be entitled to the excess.

FIFTH: *Method and Place of Delivery.* Deliveries shall be by Seller's tank wagons or trucks into storage tanks of Buyer at Buyer's place of business at

SIXTH: *Terms of Payment.*

provided, however that if credit is extended and if at any time Buyer's financial responsibility shall become impaired or unsatisfactory to Seller, advance cash payment or satisfactory security shall be made or given by Buyer on demand by Seller, which may withhold further deliveries until such payment or security is received, or on any failure to receive same, Seller may at its option terminate this contract. Discounts for cash, if any, shall not be allowed from or applied to or computed upon taxes.

SEVENTH: *Duration of Contract.* This contract shall continue in effect for a period of one year, from........................... to...........................

EIGHTH: *Other Provisions.* The *Provisions of Contract* printed on the reverse side, together with such other provisions as are contained in any rider attached hereto, to the extent applicable, are incorporated herein and made a part hereof. No modification of this contract shall be binding on Seller unless in writing and signed by Seller.

Executed and delivered the day and year first above written.

STANDARD OIL OF NEW YORK
Division of
SOCONY-VACUUM OIL COMPANY, INCORPORATED

Negotiated by:

By ....

....
SELLER'S SALESMAN

Standard Oil of New York, a division of Socony-Vacuum Oil Co., Inc.
Tank wagon consumers gasoline contract. $8–12

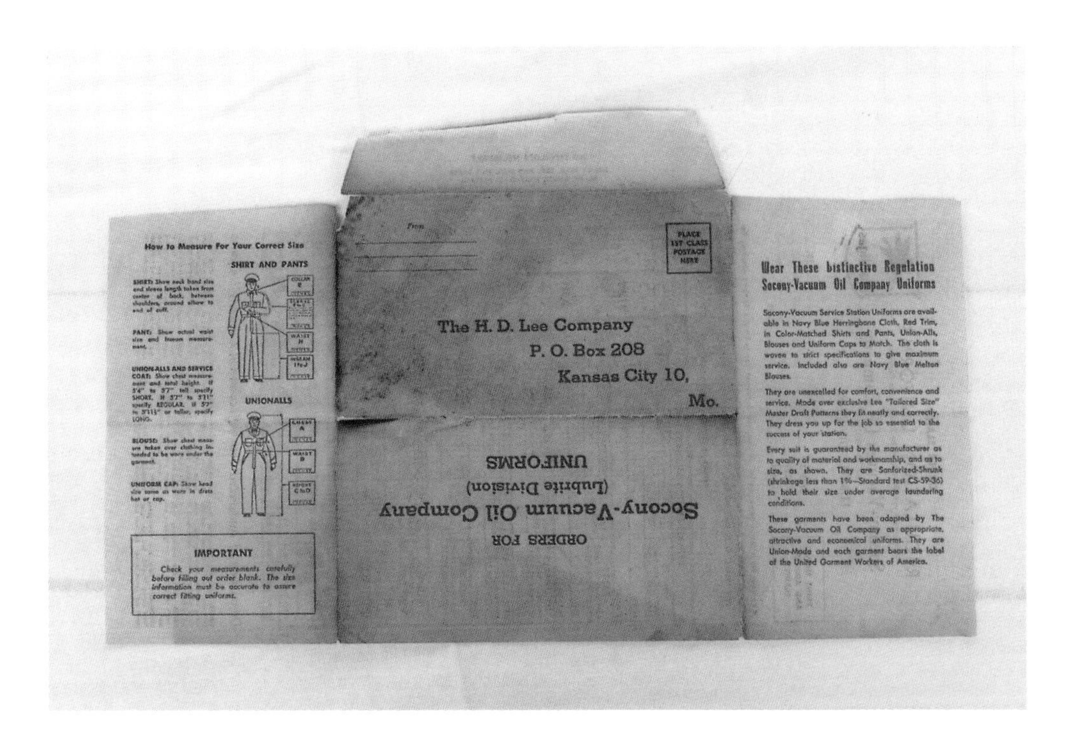

A-14

ORDER BLANK

REGULATION UNIFORMS

Union-Made and Guaranteed by The H. D. Lee Company

Socony-Vacuum Oil Company

INCORPORATED

LUBRITE DIVISION

Mobilgas

PRINT NAME AND ADDRESS PLAINLY

DATE ______ 19___

SHIP TO ______

STREET ______

TOWN ______ STATE ______

● Contract Prices Shown Below are for all sizes, with Regulation Socony-Vacuum Emblems Attached, Delivered to You by Parcel Post if Money Order or Cashier's Check Accompanies Order. If You Prefer Shipment Made C. O. D. There will Be C. O. D. and Return Money Order Charges in Addition to Prices Shown.

SUBJECT TO CHANGE WITHOUT NOTICE

SERVICE STATION SALESMEN'S UNIFORMS	LOT		PRICE EACH DELIVERED	TOTAL AMOUNT
SHIRTS — NAVY BLUE, RED PIPED, HERRINGBONE JEANS	S-1875	SHOW NECK BAND (N) AND SLEEVE LENGTH (S): N S N S N S N S N S N S — SIZE / QUANTITY	$1.82	2.70
PANTS — NAVY BLUE, RED PIPED, HERRINGBONE TWILLS	S-1778	SHOW WAIST MEASURE (W) AND INSEAM LENGTH (L): W L W L W L W L W L W L — SIZE / QUANTITY	2.30	3.45
UNION-ALLS — NAVY BLUE, RED TRIM, HERRINGBONE TWILLS	S-1258	SHOW HEIGHT HERE — SHOW NUMBER OF GARMENTS IN SIZE SQUARE: 34 36 38 40 42 44 46 48 50 — QUANTITY	3.96	5.95
SERVICE COATS — NAVY BLUE, RED TRIM, HERRINGBONE TWILLS	S-1868	SHOW HEIGHT HERE — SHOW NUMBER OF GARMENTS IN SIZE SQUARE: 34 36 38 40 42 44 46 48 50 — QUANTITY	Temporarily Discontinued	5.65
BLOUSES — NAVY BLUE, RED TRIM, HERRINGBONE TWILLS	S-1776-BS	SHOW HEIGHT HERE — SHOW NUMBER OF GARMENTS IN SIZE SQUARE: 34 36 38 40 42 44 46 48 50 — QUANTITY	Temporarily Discontinued	
BLOUSES — NAVY BLUE MELTON BUTTON FRONT	745-BB	SHOW HEIGHT HERE — SHOW NUMBER OF GARMENTS IN SIZE SQUARE: 34 36 38 40 42 44 46 48 50 — QUANTITY	Temporarily Discontinued	
UNIFORM CAPS — TWO TOPS TO MATCH ABOVE UNIFORMS, VENTILATED CANE FRAME WITH BAND	C-1389W	SHOW NUMBER OF CAPS IN PROPER HEAD SIZE SQUARE: 6⅞ 6⅞ 7 7⅛ 7¼ 7⅜ 7½ 7⅝ — QUANTITY	1.63	2.45
TOPS ONLY			[illegible]	[illegible]
BOW TIES	T-2	BLACK CREPE ADJUSTABLE — Number ___	.23	.30
STATE SALES OR USE TAX		(If tax is payable in state where you reside - add here) %		
Do not send currency unless by registered mail		TOTAL AMOUNT OF ORDER (Except C. O. D. Charges)		

NO ENVELOPE NECESSARY

SIMPLY FOLD, SEAL AND MAIL THIS ORDER BLANK UNDER FIRST-CLASS POSTAGE

Printed in U. S. A.

Description of Authorized Socony-Vacuum Oil Company Uniforms

SHIRTS: Navy Blue Herringbone Jeans. Both sides of front pleat piped in red. Red piping at top of pocket flaps and on cuffs. Flying Red Horse Emblem embroidered on left pocket.

PANTS: Navy Blue Herringbone Twills (Heavy Weight). Red piping down outseam.

UNION-ALLS: Navy Blue Herringbone Twills (Heavy Weight). Red trim collar, lapels and pocket facings. Red belt with scratch-proof safety cover buckle. Flying Red Horse Emblem embroidered on left breast pocket.

SERVICE COAT: Navy Blue Herringbone Twills (Heavy Weight). Red trim collar, lapels and pocket facings. Red belt with scratch proof safety cover buckle. Flying Red Horse Emblem embroidered on left breast pocket.

BLOUSES: Navy Blue Herringbone Twills (Heavy Weight). Band bottom, button front. Red trim collar. Flying Red Horse Emblem embroidered on left breast pocket. Navy Blue Melton, Leather Trim Talon Fastener. Front has Flying Red Horse Emblem attached.

CAPS: Ventilated Cane Frame with Regular style top, wire-stiffened snap-on tops of Navy Blue Herringbone Twills. Flying Red Horse Emblem attached to tops. Two tops and rayon cap band supplied with each cap. Suitable for both summer and winter wear.

Socony-Vacuum Oil co., Inc. Lubrite Division. Uniform order form and mailer in one (front & back). $35

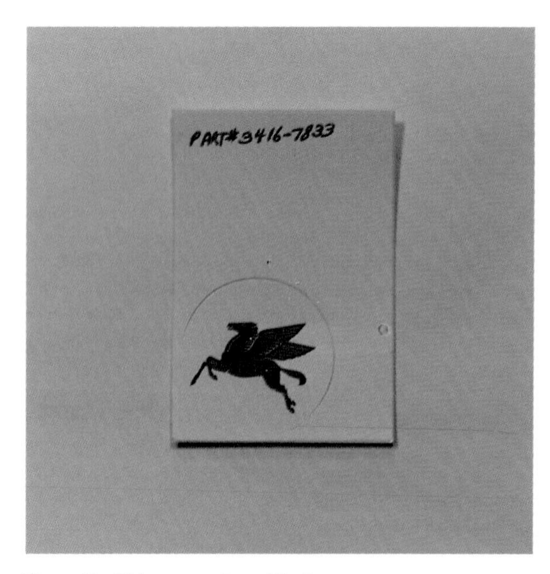

Flying Red Horse sticker. $5–8

Flying Red Horse drinking cup. $5–8

Socony-Vacuum Oil Co., Inc, White Eagle Division.
Book of Magic Tricks & Puzzles. $20–25

Socony Mobil. Terminal & Bulk plant safety manual in a vinyl cover. $15–20

Three-ring binder for Mobil Tires, Batteries and Accessories literature. $15–20

Mobil Customer Service Record box. $175

Mobil Dealers Manual and an interior page. $35–50

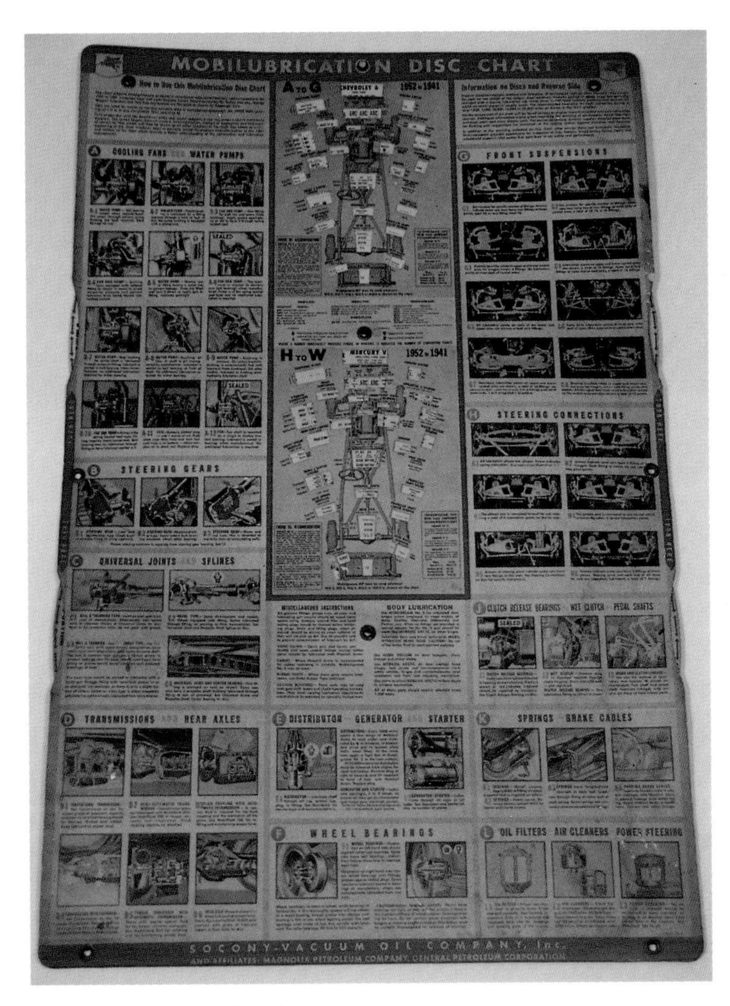

Socony-Vacuum Oil Co., Mobil lubrication chart for vehicles from 1941 to 1952. $50–65

Socony Mobil. Mobil lubrication chart for vehicles from 1949 to 1957. $50–65

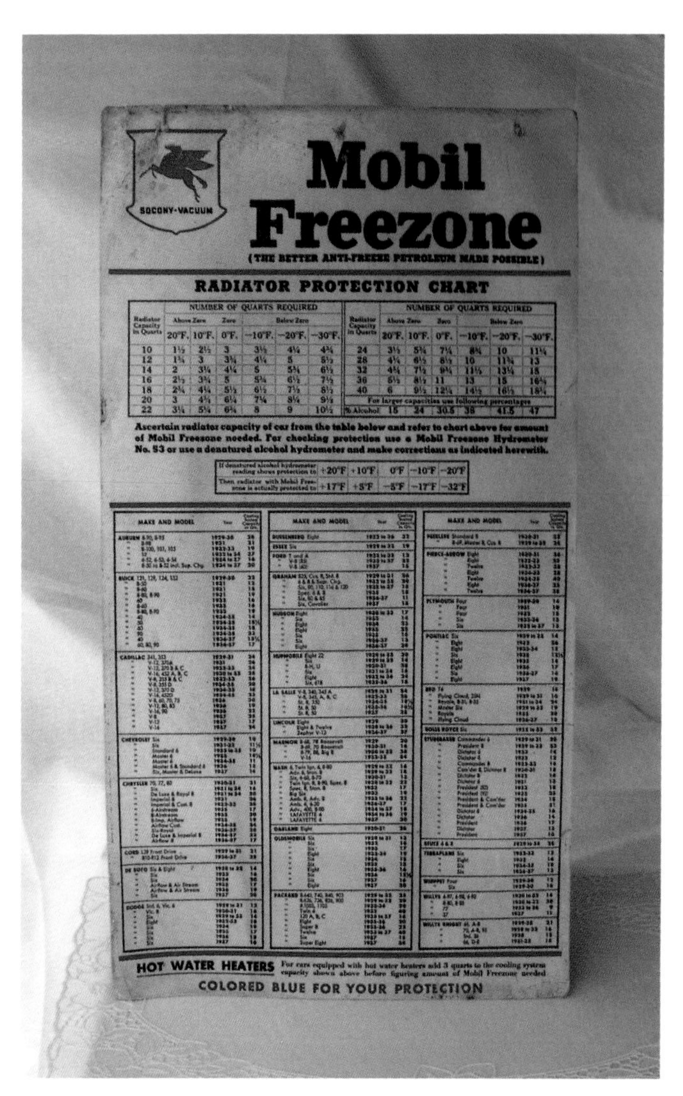

Socony-Vacuum. Mobil Freezone radiator chart. $15–25

1961 Lubrication recommendation chart for a service station. $45–60

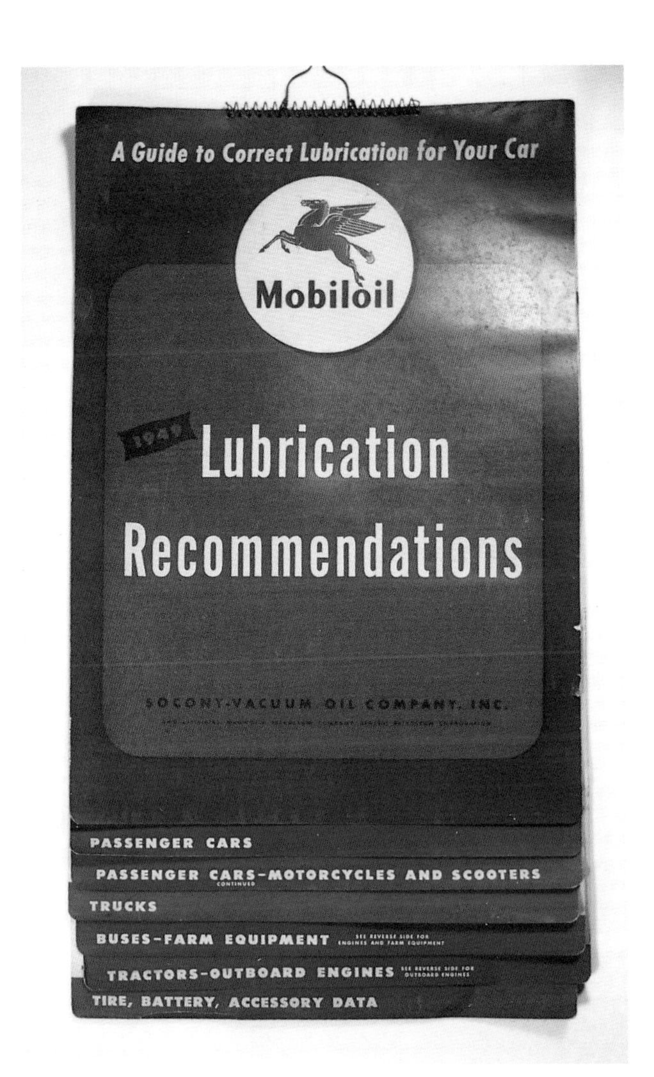

1949 Lubrication recommendation chart for a service station. $55–70

Blank matchbook cover sheet. $20–30

A variety of matchbook covers. $8–20

IX. Collections

George Deel's hat, featuring a Magnolia gasoline badge pin.

A portion of one collector's collection.

Authorized Service
VACUUM OIL COMPANY
GARGOYLE
Mobiloil
GASOLINE
SOCONY
Ford Cars
Gargoyle Mobiloil
The Chart Specifies
The Correct Grades to
Lubricate Your Engine
SANILAC
Axle Grease
Mobiloil Arctic

GARGOYLE
Mobiloil
Make the chart your guide
Mobiloil
"A"
Mobiloil
"BB"
Mobiloil
Arctic

Another collection.

Bibliography

Anderton, Mark and Sherry Mullen. *Gas Station Collectibles.* Wallace-Homestead Book Company, Radnor, Pennsylvania, 1994

Brunner, Mike. *Gasoline Treasures with Values.* Schiffer Publishing, LTD, Atglen, Pennsylvania, 1996

Gold Rush Dazes, Rochester, Minnesota, 1996, 1997 and 1998

Iowa Gas Swap Meet, August 1997

Mobil Corporation. *Mobil at 125,* Mobil Corporation, Fairfax, Virginia

Mobil Corporation. *A Brief History of Mobil,* Mobil Corporation, Fairfax, Virginia

Northland Auction Service, Delano, Minnesota, Spring 1996

Pease, Rick. *Filling Station Collectibles with Price Guide.* Schiffer Publisher, LTD., Atglen Pennsylvania, 1994.

Pease, Rick. *Petroleum Collectibles with Prices.* Schiffer Publishing, LTD., Atglen, Pennsylvania, 1997.

Stenzler, Mitch and Rich Pease. *Gas Station Collectibles with Price Guide.* Schiffer Publishing, LTD., Atglen, Pennsylvania, 1993.